Programmable
Logic Controllers

Programmable Logic Controllers

Programming Methods and Applications

John R. Hackworth
Old Dominion University

Frederick D. Hackworth, Jr.

PEARSON

Prentice
Hall

Upper Saddle River, New Jersey
Columbus, Ohio

Library of Congress Cataloging-in-Publication Data

Hackworth, John R.
   Programmable logic controllers : programming methods and applications / by John R.
Hackworth and Frederick D. Hackworth, Jr.
      p. cm.
   Includes bibliographical references and index.
   ISBN 0-13-060718-5
      1. Programmable controllers. I. Hackworth, Frederick D. II. Title.

TJ223.P76.H33 2004
629.8'9—dc21   2003043618

Editor in Chief: Stephen Helba
Assistant Vice President and Publisher: Charles E. Stewart, Jr.
Assistant Editor: Mayda Bosco
Production Editor: Alexandrina Benedicto Wolf
Production Coordination: Carlisle Publishers Services
Design Coordinator: Diane Ernsberger
Cover Designer: Linda Sorrells-Smith
Cover art: Digital Vision
Production Manager: Matt Ottenweller
Marketing Manager: Ben Leonard

Pearson Prentice Hall™ is a trademark of Pearson Education, Inc.
Pearson® is a registered trademark of Pearson plc
Prentice Hall® is a registered trademark of Pearson Education, Inc.

Pearson Education Ltd.   Pearson Education Australia Pty. Limited
Pearson Education Singapore Pte. Ltd.   Pearson Education North Asia Ltd.
Pearson Education Canada, Ltd.   Pearson Educación de Mexico, S.A. de C.V.
Pearson Education—Japan   Pearson Education Malaysia Pte. Ltd.

ISBN 0-13-060718-5

Dedicated to Linda and Phyllis,
and to our Dad,
who introduced us to
electrical engineering.

Preface

Most textbooks related to programmable logic controllers (PLCs) start with the basics of ladder logic, Boolean algebra, contacts, coils, and all the other aspects of learning to program PLCs. However, once they get more deeply into the subject, these books generally narrow the field of view to one particular manufacturer's unit (usually one of the more popular brands and models) and concentrate on programming that device with its capabilities and peculiarities. This is worthwhile if the desire is simply to learn to program that particular unit. However, after finishing the PLC course, most students will likely be employed designing, programming, and maintaining systems using PLCs of another brand or model or various machines with different PLC brands and models. We believe that it is more advantageous to approach the study of PLCs using a general language that provides a thorough knowledge of programming concepts which can be adapted to all controllers. This language would be based on a collection of different manufacturer types with generally the same programming technique and capability. Although it would be impossible to teach one programming language and technique that would be applicable to each and every programmable controller on the market, students can be given a thorough insight into programming methods with this general approach that will allow them to easily adapt to any PLC situation encountered.

The goal of this text is to help the reader develop a good general working knowledge of programmable controllers while concentrating on relay ladder logic techniques and how the PLC is connected to external components in an operating control system. The text presents real-world programming problems that can be solved on any available programmable controller or PLC simulator. Later chapters relate to more advanced subjects that are more suitable for an advanced course in machine controls.

Readers should have a thorough understanding of fundamental ac and dc circuits, electronic devices (including thyristors) and a knowledge of basic logic gates, flip flops, Boolean algebra, and college algebra and trigonometry. Although a knowledge of calculus will enhance the understanding of closed-loop controls, it is not required.

We also hope that this text will serve as a technical reference for students and professionals.

Acknowledgments

We would like to thank the following reviewers for their valuable feedback: J. Don Book, Pittsburg State University, Kansas; Bob Jackson, DeVry Institute of Technology, Georgia; Roger A. Kuntz, Pennsylvania State University at Erie; David J. Malooley, Indiana State University; and Rick Miller, Ferris State University, Michigan.

Contents

CHAPTER 1   LADDER DIAGRAM FUNDAMENTALS   1

Objectives   1
Introduction   1
1–1   Basic Components and Their Symbols   2
1–2   Fundamentals of Ladder Diagrams   13
1–3   Machine Control Terminology   31
Summary   33

CHAPTER 2   THE PROGRAMMABLE LOGIC CONTROLLER   35

Objectives   35
Introduction   35
2–1   A Brief History   36
2–2   PLC Configurations   37
2–3   System Block Diagram   42
2–4   Update-Solve the Ladder-Update   43
2–5   Update   44
2–6   Solve the Ladder   45
Summary   48

CHAPTER 3   FUNDAMENTAL PLC PROGRAMMING   50

Objectives   50
Introduction   50
3–1   Physical Components vs. Program Components   51
3–2   Example Problem—Lighting Control   56
3–3   Internal Relays   58
3–4   Disagreement Circuit   59

3–5   Majority Circuit   59
3–6   Oscillator   62
3–7   Holding (also called Sealed, or Latched) Contacts   64
3–8   Always-ON and Always-OFF Contacts   65
3–9   Ladder Diagrams Having More Than One Rung   67
Summary   69

CHAPTER 4   ADVANCED PROGRAMMING TECHNIQUES   72

Objectives   72
Introduction   72
4–1   Ladder Program Execution Sequence   73
4–2   Flip Flops   73
4–3   R-S Flip Flop   73
4–4   One Shot   74
4–5   D Flip Flop   77
4–6   T Flip Flop   79
4–7   J-K Flip Flop   81
4–8   Counters   83
4–9   Sequencers   85
4–10   Timers   87
4–11   Master Control Relays and Control Zones   96
Summary   98

CHAPTER 5   MNEMONIC PROGRAMMING CODE   103

Objectives   103
Introduction   103
5–1   AND Ladder Rung   104
5–2   Entering Normally Closed Contacts   105
5–3   OR Ladder Rung   106
5–4   Simple Branches   107
5–5   Complex Branches   110
Summary   112

CHAPTER 6   WIRING TECHNIQUES   115

Objectives   115
Introduction   115
6–1   PLC Power Connection   116

6–2   Input Wiring   119
6–3   Inputs Having a Single Common   121
6–4   Isolated Inputs   124
6–5   Output Wiring   126
6–6   Relay Outputs   127
6–7   Solid State Outputs   131
Summary   137

CHAPTER 7   ANALOG I/O   139

Objectives   139
Introduction   139
7–1   Analog (A/D) Input   110
7–2   Analog (D/A) Output   144
7–3   Analog Data Handling   145
7–4   Analog Input Potential Problems   146
Summary   147

CHAPTER 8   DISCRETE POSITION SENSORS   150

Objectives   150
Introduction   150
8–1   Sensor Output Classification   151
8–2   Connecting Discrete Sensors to PLC Inputs   154
8–3   Proximity Sensors   156
8–4   Inductive Proximity Sensors   156
8–5   Capacitive Proximity Sensors   159
8–6   Ultrasonic Proximity Sensors   161
8–7   Optical Proximity Sensors   163
Summary   167

CHAPTER 9   ENCODERS, TRANSDUCERS, AND ADVANCED SENSORS   169

Objectives   169
Introduction   169
9–1   Temperature   170
9–2   Liquid Level   175
9–3   Force   177
9–4   Pressure/Vacuum   181
9–5   Flow   186

9–6   Inclination   192
9–7   Acceleration   193
9–8   Angle Position Sensors   194
9–9   Linear Displacement   204
Summary   209

CHAPTER 10   CLOSED-LOOP AND PID CONTROL   212

Objectives   212
Introduction   212
10–1   Simple Closed-Loop Systems   212
10–2   Problems with Simple Closed-Loop Systems   214
10–3   Closed-Loop Systems Using Proportional, Integral, Derivative (PID)   218
10–4   Derivative Function   220
10–5   Integral Function   225
10–6   The PID in Programmable Logic Controllers   228
10–7   Tuning the PID   229
10–8   The "Adjust and Observe" Tuning Method   231
10–9   The Ziegler-Nichols Tuning Method   233
10–10   Autotuning PID Systems   239
Summary   240

CHAPTER 11   MOTOR CONTROLS   242

Objectives   242
Introduction   242
11–1   AC Motor Starter   243
11–2   AC Motor Overload Protection   245
11–3   Specifying a Motor Starter   247
11–4   DC Motor Controller   248
11–5   Variable Speed (Variable Frequency) AC Motor Drive   255
Summary   258

CHAPTER 12   SYSTEM INTEGRITY AND SAFETY 260

Objectives   260
Introduction   260
12–1   System Integrity   260
12–2   Equipment Temperature Considerations   263
12–3   Fail Safe Wiring and Programming   264
12–4   Safety Interlocks   268
Summary   272

APPENDIX A LOGIC SYMBOLS   274

APPENDIX B INDUSTRIAL ELECTRICAL SYMBOLS   276

BIBLIOGRAPHY   281

GLOSSARY   283

INDEX   297

Ladder Diagram Fundamentals

Objectives

Upon completion of this chapter, you will know:

- the parts of an electrical machine control diagram including rungs, branches, rails, contacts, and loads.

- how to correctly design and draw a simple electrical machine control diagram.

- the difference between an electronic diagram and an electrical machine diagram.

- the diagramming symbols for common components such as switches, control transformers, relays, fuses, and time delay relays.

- the more common machine control terminology.

Introduction

Machine control design is a unique area of engineering that requires the knowledge of certain specific and unique diagramming techniques called ladder diagramming. Although there are similarities between control diagrams and electronic diagrams, many of the component symbols and layout formats are different. This chapter provides a study of the fundamentals of developing, drawing, and understanding ladder diagrams. We will begin with a description of some of the fundamental components used in ladder diagrams. The basic symbols will then be used in a study of boolean logic as applied to relay diagrams. More complicated circuits will then be discussed.

1-1 Basic Components and Their Symbols

We shall begin with a study of the fundamental components used in electrical machine controls and their ladder diagram symbols. It is important to understand that the material covered in this chapter is by no means a comprehensive coverage of all types of machine control components. Instead, we will discuss only those most commonly used. Some of the more exotic components will be covered in later chapters.

Control Transformers

For safety reasons, machine controls are low voltage components. Because the switches, lights, and other components must be touched by operators and maintenance personnel, it is contrary to electrical code in the United States to apply a voltage higher than 120 VAC to the terminals of any operator controls. For example, assume a maintenance person is changing a burned-out indicator lamp on a control panel and the lamp is powered by 480 VAC. If the person were to touch any part of the metal bulb base while it is in contact with the socket, the shock could be lethal. However, if the bulb is powered by 120 VAC or less, the resulting shock would likely be much less severe.

In order to make large, powerful machines efficient and cost effective and reduce line current, most are powered by high voltages (240 VAC, 480 VAC, or more). This means the line voltage must be reduced to 120 VAC or less for the controls. This is done using a **control transformer.** Figure 1–1 shows the electrical diagram symbol for a control transformer. The most obvious peculiarity here is that the symbol is rotated 90° with the primaries on top and secondary on the bottom. As will be seen later, this is done to make it easier to draw the remainder of the ladder diagram. Notice that the transformer has two primary windings. These

FIGURE 1–1
Control Transformer

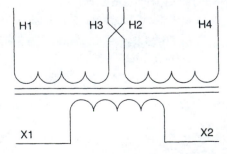

are usually rated at 240 VAC each. By connecting them in parallel (strapping H1 to H3 and H2 to H4), we obtain a 240 VAC primary, and by connecting them in series (strapping H2 to H3), we have a 480 VAC primary. The secondary windings are generally rated at 120 VAC, 48 VAC, or 24 VAC. By offering control transformers with dual primaries, transformer manufacturers can reduce the number of transformer types in their product line and make their transformers more versatile and less expensive.

Fuses

Control circuits are always fuse protected. This prevents damage to the control transformer in the event of a short in the control circuitry. The electrical symbol for a fuse is shown in Figure 1–2. The fuse used in control circuits is generally a slo-blow fuse (i.e., it is generally immune to current transients that occur when power is switched on). It must be rated at a current that is less than or equal to the rated secondary current of the control transformer, and it must be connected in series with the transformer secondary.

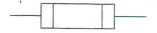

FIGURE 1–2
Fuse

Most control transformers can be purchased with a fuse block (fuse holder) for the secondary fuse mounted on the transformer, as shown in Figure 1–3.

FIGURE 1–3
*Control
Transformer with
Secondary Fuse
Holder*

Switches

There are two fundamental uses for switches. First, switches are used for operator input to send instructions to the control circuit. Second, switches may be installed on the moving parts of a machine to provide automatic feedback to the control system. There are many different types of switches—too many to cover in this text. However, with a basic understanding of switches, it is easy to understand most of the different types.

Pushbutton   The most common switch is the pushbutton. It is also the one that needs the least description because it is widely used in automotive and electronic equipment applications. There are two types of pushbuttons—the momentary and maintained. The **momentary pushbutton** switch is activated when the button is pressed and deactivated when the button is released. The deactivation is done using an internal spring. The **maintained pushbutton** switch activates when pressed, but remains activated when it is released. To deactivate it, it must be pressed a second time. For this reason, this type of switch is sometimes called a push-push switch. The on/off switches on most desktop computers and laboratory oscilloscopes are maintained pushbuttons.

The contacts on switches can be of two types. These are normally open (N/O) and normally closed (N/C). Whenever a switch is in its deactivated position, the N/O contacts will be open (non-conducting) and the N/C contacts will be closed (conducting). Figure 1–4 shows the schematic symbols for (a) a normally open pushbutton and (b) a normally closed pushbutton. The symbol of Figure 1–4c is a single pushbutton with both N/O and N/C contacts. There is no internal electrical connection between different contact pairs on the same switch. Most industrial switches can have extra contacts "piggy backed" on the switch, so as many contacts as needed of either type can be added by the designer.

FIGURE 1–4

Momentary Pushbutton Switches

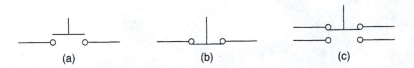

(a)   (b)   (c)

The schematic symbol for the maintained pushbutton is shown in Figure 1–5. Note that it is the symbol for the momentary pushbutton with a "see-saw" mechanism added to hold in the switch actuator until it is pressed a second time. As with the momentary switch, the maintained switch can have as many contacts of either type as desired.

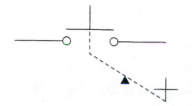

FIGURE 1–5
*Maintained
Switch*

Pushbutton Switch Actuators   The actuator of a pushbutton is the part of the switch that is depressed to activate the switch. These actuators are available in several different styles as shown in Figure 1–6. Each has a specific purpose.

The switch in Figure 1–6a has a **guarded** or **shrouded actuator.** In this case, the pushbutton is recessed ¼″ to ½″ inside the sleeve and can only be depressed by an object smaller than the sleeve (such as a finger). It provides protection against the button being accidentally depressed by the palm of the hand or other object and is therefore used in situations where pressing the switch causes something potentially dangerous to happen. Guarded pushbuttons are used in applications such as START, RUN, CYCLE, JOG, or RESET operations. For example, the RESET pushbutton on your computer is likely a guarded pushbutton.

The switch shown in Figure 1–6b is called a **flush pushbutton.** It has an actuator that is aligned to be even with the sleeve. It provides protection against accidental actuation similar to the guarded pushbutton; however, since it is not

(a)   (b)   (c)

FIGURE 1–6
Switch Actuators—(a) Guarded, (b) Flush, and (c) Extended (Courtesy of IDEC Corporation)

recessed, the level of protection is not to the extent of the guarded pushbutton. This type of switch actuator works better in applications where it is desired to back light the actuator (called a **lighted pushbutton**).

The switch in Figure 1–6c is an **extended pushbutton.** Obviously, the actuator extends beyond the sleeve, which makes the button easy to depress by finger, palm of the hand, or any other object. It is intended for applications where it is desirable to make the switch as accessible as possible such as STOP, PAUSE, or BRAKES.

The three types of switch actuators shown in Figure 1–6 are not generally used for applications required in emergency situations nor for operations that occur hundreds of times per day. For both of these applications, a switch is needed that is the most accessible of all switches. These types are the **mushroom head** or **palm head pushbutton** (sometimes called **palm switch,** for short), and are illustrated in Figure 1–7.

Although these two applications are radically different, the switches look similar. The mushroom head switch shown in Figure 1–7a is a momentary switch that may cause a machine to run one cycle of an operation. For safety reasons, they are usually used in pairs, separated by about 24", and wired so that they must both be pressed at the same time in order to cause the desired operation to commence. When arranged and wired this way, we create what is called a **two-handed palming operation.** By doing so, we know that when the machine is cycled, the operator has both hands on the pushbuttons and not in the machine.

The switch in Figure 1–7b is a detent pushbutton (i.e., when pressed in it remains in, and then to return it to its original position, it must be pulled out) and is called an **Emergency Stop,** or **E-Stop** switch. The mushroom head is always red, and

FIGURE 1–7

Mushroom Head Pushbuttons— (a) Palm and (b) Emergency Stop (Courtesy of IDEC Corporation)

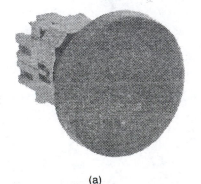

(a)   (b)

the switch is used to shut off power to the controls of a machine when the switch is pressed in. In order to restart a machine, the E-Stop switch must be pulled to the out position to apply power to the controls before attempting to run the machine.

Mushroom head switches have special schematic symbols as shown in Figure 1–8. Notice that they are drawn as standard pushbutton switches but have a curved line on the top of the actuators to indicate that the actuators have a mushroom head.

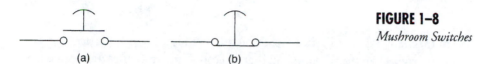

(a)   (b)

FIGURE 1–8
Mushroom Switches

Selector Switches   A selector switch is also known as a rotary switch. An automobile ignition switch and an oscilloscope's vertical gain and horizontal timebase switches are examples of selector switches. Selector switches use the same symbol as a momentary pushbutton, except a lever is added to the top of the actuator as shown in Figure 1–9. The switch in Figure 1–9a is open when the selector is turned to the left and closed when turned to the right. The switch in Figure 1–9b has two sets of contacts. The top contacts are closed when the switch selector is turned to the left position and open when the selector is turned to the right. The bottom set of contacts works exactly opposite. There is no electrical connection between the top and bottom pairs of contacts. In most cases, we label the selector positions the same as the labeling on the panel where the switch is located. For the switch in Figure 1–9b, the control panel would be labeled with the STOP position to the left and the RUN position to the right.

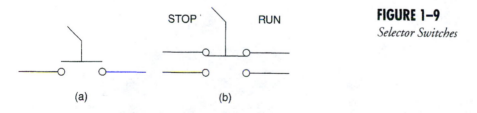

(a)   (b)

FIGURE 1–9
Selector Switches

Limit Switches   Normally, limit switches are not operator accessible. Instead, they are activated by moving parts on the machine. They are usually mechanical switches but can also be light activated (such as the automatic door openers used by stores and supermarkets) or magnetically operated (such as the magnetic switches used on home security systems that sense when a window has been opened). An example of a mechanically operated limit switch is the switch on the refrigerator door that turns on the light inside. They are sometimes called cam switches because many are operated by a camming action when a moving part passes by the switch. The symbols for both types of limit switches are shown in Figure 1–10. The N/O version is in Figure 1–10a, and the N/C version is in Figure 1–10b.

FIGURE 1–10
Limit Switches

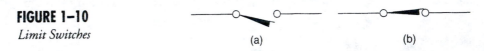

(a)   (b)

One of the many types of limit switch is pictured in Figure 1–11. Notice that this switch has a wheel on the actuator arm that can roll along a moving surface or edge.

Indicator Lamps

All control panels include indicator lamps. They tell the operator when power is applied to the machine and indicate the present operating status of the machine. In a schematic diagram, an indicator is drawn as a circle with "light rays" extending on the diagonals as shown in Figure 1–12.

Although the light bulbs used in indicators are generally incandescent (white), they are usually covered with colored lenses. The colors are usually red, green, or amber, but other colors are also available. Red lamps are reserved for safety critical indicators (power is on, the machine is running, an access panel is open, or that a fault has occurred). Green usually indicates safe conditions (power to the motor is off, brakes are on, etc.). Amber indicates conditions that are important but not dangerous

FIGURE 1–11
Limit Switch

FIGURE 1–12
Lamp

(fluid getting low, machine paused, machine warming up, etc.). Other colors indicate information not critical to the safe operation of the machine (time for preventive maintenance, etc.). Sometimes it is important to attract the operator's attention with a lamp. In these cases, we usually flash the lamp continuously on and off.

Relays

Early electrical control systems were mainly composed of relays and switches. Although switches are well-known devices, that may not be the case for relays. Therefore, before continuing our discussion of machine control ladder diagramming, a brief discussion of relay fundamentals may be beneficial. A simplified, cutaway (cross section) drawing of a relay with one contact set is shown in Figure 1–13.

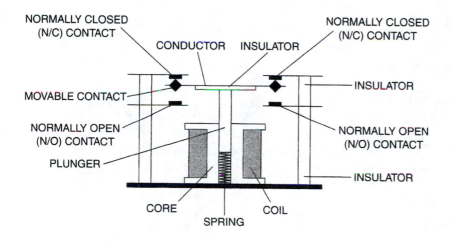

FIGURE 1–13
Relay or Contactor

A relay, or **contactor**, is an electromagnetic device composed of a frame (or core), an electromagnet coil, and contacts (some movable and some fixed). The movable contacts (and the conductor that connects them) are mounted via an insulator to a plunger that moves within a bobbin. A coil of copper wire is wound on the bobbin to create an electromagnet. A spring holds the plunger up and away from the electromagnet. When the electromagnet is energized by passing an electric current through the coil, the magnetic field pulls the plunger into the core, which pulls the movable contacts downward. Two fixed pairs of contacts are mounted to the relay frame on electrical insulators so that when the movable contacts are not being pulled toward the core (the coil is de-energized) they physically touch the upper fixed pair of contacts and, when being pulled toward the coil, touch the lower pair of fixed contacts. There can be several sets of contacts mounted to the relay frame. The contacts energize and de-energize as a result of applying power to the relay coil (connections to the relay coil are not shown). Referring to Figure 1–13, when the coil is de-energized, the movable contacts are connected to the upper fixed contact pair. These fixed contacts are referred to as the **normally closed contacts** because they are bridged together by the movable contacts and conductor whenever the relay is in its "power off" state. Likewise, the movable contacts are *not* connected to the lower fixed contact pair when the relay coil is de-energized. These fixed contacts are referred to as the **normally open contacts.** *Contacts are named with the relay in the de-energized state.* Normally open contacts are said to be **OFF** when the coil is de-energized and **ON** when the coil is energized. Normally closed contacts are ON when the coil is de-energized and OFF when the coil is energized. Generally, designers who are familiar with digital logic tend to think of N/O contacts as non-inverting contacts and N/C contacts as inverting contacts.

It is important to remember that many of the schematic symbols used in electrical diagrams are different than the symbols for the same types of components in electronic diagrams. Figure 1–14 shows the three most common relay symbols used in electrical machine diagrams. These three symbols are (a) normally open contact, (b) normally closed contact, and (c) coil. Notice that the normally open contact on the left could easily be misconstrued as a capacitor by an electronic designer. That is why it is important when working with electrical machines to mentally "shift gears" to think in terms of electrical symbols and not electronic symbols.

FIGURE 1–14

Relay Symbols

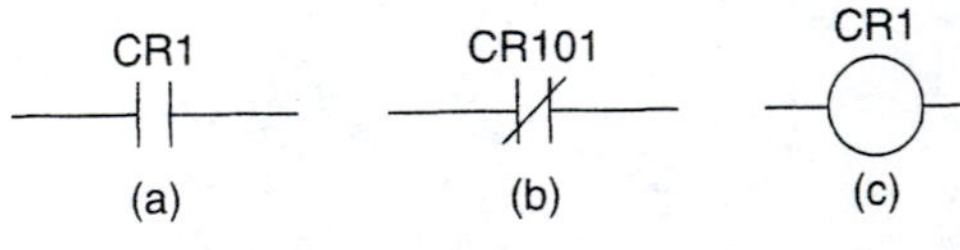

CR1   CR101   CR1

(a)   (b)   (c)

Notice that the normally closed and normally open contacts of Figure 1–14 each have lines extending from both sides of the symbol. These are the connection lines which, on a real relay, would be the connection points for wires. The reader is invited to refer back to Figure 1–13 and identify the relationship between the normally open and normally closed contacts on the physical relay and their corresponding symbols in Figure 1–14.

The coil symbol shown in Figure 1–14 represents the coil of the relay we have been discussing. The coil, like the contacts, has two connection lines extending from either side. These represent the physical wire connections to the coil on the actual relay. Notice that the coil and contacts in the figure each have a reference designator label above the symbol. This label identifies the contact or coil within the diagram. Coil CR1 is the coil of relay CR1. When coil CR1 is energized, all the normally open CR1 contacts will be closed, and all the normally closed CR1 contacts will be open. Likewise, if coil CR1 is de-energized, all the normally open CR1 contacts will be open and all the normally closed CR1 contacts will be closed. Most coils and contacts we use will be labeled as CR (CR is the abbreviation for "control relay"). A contact labeled CR indicates that it is associated with a relay coil. Each relay will have a specific number associated with it. The range of numbers used will depend upon the number of relays in the system.

Figure 1–15 shows the same relay symbols as in Figure 1–14, however, they have not been drawn graphically. Instead, they are drawn using standard ASCII printer characters (hyphens, vertical bars, forward slashes, and parentheses). This is a common method used when the diagram is generated by a computer on an older printer, or when it is desired to rapidly print the diagram (ASCII characters print very quickly). This printing method is usually limited to diagrams of PLC programs (as we will see later). Machine electrical diagrams are rarely drawn using this method.

```
    CR1              CR101            CR1
----| |----       ----|/|----      ---- ( ) ----
    (a)              (b)              (c)
```

FIGURE 1–15

ASCII Relay Symbols

Relays are available in an extremely wide range of sizes and current and voltage capacities. Small reed relays in 14 pin DIP integrated circuit packages are capable of switching a few tenths of an ampere at less than 100 volts, while large contactors the size of a room can be capable of switching thousands of amperes at thousands of volts. However, for electrical machine diagrams, the schematic symbol for a relay is the same regardless of the relay's size.

Time Delay Relays

It is possible to construct a relay with a built-in time delay device that causes the relay contacts to switch ON or switch OFF after a time delay. These types of relays are called **time delay relays,** or TDRs. The schematic symbols for a TDR coil and contacts are the same as for a conventional control relay except that the coil symbol has the letters "TDR" or "TR" written inside or next to the coil symbol. The relay itself looks similar to any other relay except that it has a control knob that allows the user to set the amount of time delay. There are two basic types of time delay relay—delay-on timer, sometimes called a TON (pronounced Tee-ON), and the delay-off timer, sometimes called a TOF (pronounced Tee-Off). It is important to understand the difference between these relays in order to specify and apply them correctly.

Delay-On Timer (TON) Relay   When an on-timer is installed in a circuit, the user adjusts the control on the relay for the desired time delay. This time setting is called the **preset.** Figure 1–16 shows a timing diagram of a **delay-on** time delay relay, and to the right of each waveform, it shows the schematic symbol for each component of the relay. Notice on the top waveform that we are simply turning on power to the relay's coil and, at some undetermined time, turning it off (the amount of time that the coil is energized makes no difference to the operation of the relay). When the coil is energized, the internal timer in the relay begins running (this can be either a motor-driven mechanical timer or an electronic timer). When the time value contained in the timer reaches the preset value, the relay energizes. When this happens, all normally open (N/O) contacts on the relay close and all normally closed (N/C) contacts on the relay open. Notice also that when power is removed from the relay coil, the contacts immediately return to their de-energized state, the timer is reset, and the relay is ready to begin timing again the next time power is applied. If power is applied to the coil and then switched OFF before the preset time is reached, the relay contacts never activate.

FIGURE 1–16
Delay-On Time Delay Relay Timing Diagram and Schematic Symbols

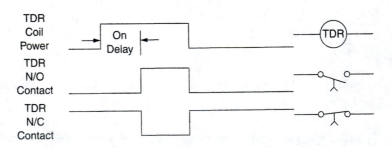

Delay-on relays are useful for delaying turn-on events. For example, when the motor is started on a machine, a TON time delay relay can be used to disable all the other controls for a few seconds until the motor has had time to achieve running speed.

Delay-Off Timer (TOF) Relay   Figure 1–17 shows a timing diagram for a **delay-off** timer. To the right of each waveform, it shows the schematic symbol for each component of the relay. In this case, at the instant power is applied to the relay coil, the contacts activate—that is, the N/O contacts close and the N/C contacts open. The time delay occurs when the relay is switched OFF. After power is removed from the relay coil, the contacts stay activated until the relay times-out. If the relay coil is re-energized before the relay times-out, the timer will reset, and the relay will remain energized until after power is removed, at which time it will again start the delay-off cycle.

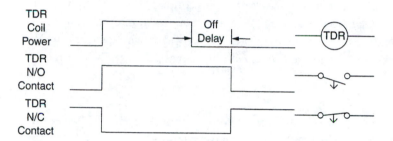

FIGURE 1–17
Delay-Off Time Delay Relay Timing Diagram and Schematic Symbols

Delay-off time delay relays are excellent for applications requiring time to be "stretched." As an example, it can be used to operate a fan that continues to cool the machine even after the machine has been stopped. TOF time delay relays are also used in outdoor lighting control motion sensors. Whenever motion in sensed, the relay immediately switches ON; when motion is no longer sensed, the TDR keeps the lights on for some period of time before de-energizing.

1–2 Fundamentals of Ladder Diagrams

Basic Diagram Framework

All electrical machine diagrams are drawn using a standard format called the **ladder diagram.** Beginning with the control transformer, we add a protective fuse on the left side. As mentioned earlier, the fuse is often part of the transformer itself. From the transformer/fuse combination, horizontal lines are extended to both sides and then drawn vertically down the page as shown in Figure 1–18. These vertical lines are called **power rails** or simply **rails** or **uprights.** The voltage difference between the two rails is equal to the transformer secondary voltage; therefore, any component connected between the two rails will be powered.

FIGURE 1–18
*Basic Control
Circuit*

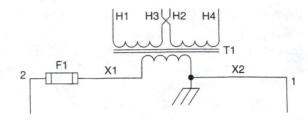

Notice that the right side of the control transformer secondary is grounded to the frame of the machine (earth ground). The reason is that without this ground, should the transformer short internally from primary to secondary, it could apply potentially lethal line voltages to the controls. With the ground, an internal transformer short will cause a fuse to blow or circuit breaker to trip farther "upstream" on the line voltage side of the transformer, which will shutdown power to the controls.

Wiring

All wires in a control system are numbered. In our diagram, the left rail is wire number 2 and the right rail is wire number 1. When the system is constructed, the actual wires used to connect the components will have a label on each end (called a **wire marker**), as shown in Figure 1–19, indicating the same wire number. This makes it easier to build, troubleshoot, and modify the circuitry. In addition, by using wire markers, all the wires will be identified, which makes it unnecessary to use more than one color wire to wire the system (except for ground wires) and reduces the cost to

FIGURE 1–19
Wire Markers

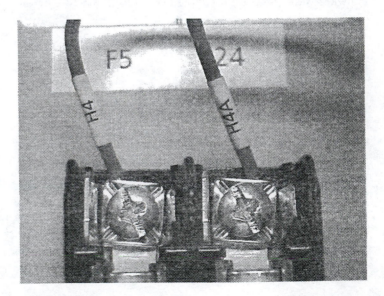

construct the machine. Generally, control circuits are wired with all black, red, or white wire (do not use green—it is reserved for safety ground wiring). Notice that in Figure 1–18 the wire connecting T1 to F1 is not numbered. This is because in our design we will be using a transformer with the fuse block included. Therefore, this will be a permanent metal strap on the transformer and will not be a wire.

The wire generally used within the control's circuitry is AWG14 or AWG16 stranded copper, type MTW or THHN. MTW is an abbreviation for "machine tool wire," and THHN indicates thermoplastic heat-resistant nylon-coated. MTW has a single PVC insulation jacket and is used in applications where the wire will not be exposed to gasoline or oil. It is less expensive, more flexible, and easier to route, bundle, and pull through conduits. THHN is used in areas where the wire may be exposed to gasoline or oil (such as hydraulically operated machines). It has a transparent, oil-resistant nylon coating on the outside of the insulation. The disadvantage to THHN is that it is more expensive, more difficult to route around corners, and its larger diameter reduces the maximum number of conductors that can be pulled into restricted space (such as inside conduits). Since most control components use low currents, AWG14 or AWG16 wire is much larger than needed. However, it is generally accepted for panel and controls wiring because the larger wire is tough, more flexible, easier to install, and can better withstand the constant vibration created by heavy machinery.

Reference Designators

For all electrical diagrams, every component is given a reference designator. This is a label assigned to the component so that it can be easily located. The reference designator for each component appears on the schematic diagram, the mechanical layout diagram, the parts list, and sometimes is even stamped on the actual component itself. The reference designator consists of an alphabetical prefix followed by a number. The prefix identifies what kind of part it is (control relay, transformer, limit switch, etc.), and the number indicates which particular part it is. Some of the most commonly used reference designator prefixes are listed here.

T	transformer
CR	control relay
R	resistor
C	capacitor
LS	limit switch
PB	pushbutton

S	switch
SS	selector switch
TDR or TR	time delay relay
M	motor or motor relay
L	indicator lamp or line phase
F	fuse
CB	circuit breaker
OL	overload switch or overload contact

The number of the reference designator is assigned by the designer beginning with the number 1. For example, control relays are numbered CR1, CR2, and so on, while fuses are F1, F2, and so on. It is generally a courtesy of the designer to state on the electrical drawing the "Last Used Reference Designators." This is done so that anyone who is assigned the job of later modifying the machine will know where to "pick up" in the numbering scheme for any added components. For example, if the drawing stated "Last Used Reference Designators:. CR15, T2, F3," then in a modification that adds a control relay, the added relay would be assigned the next sequential reference designator, CR16. This eliminates the possibility of skipping a number or having duplicate numbers. Also, if components are deleted as part of a modification, it is a courtesy to add a line of text to the drawing stating "Unused Reference Designators." This prevents someone who is reading the drawing from wasting time searching for a component that no longer exists.

Some automation equipment and machine tool manufacturers use a reversed component numbering scheme that starts with the number and ends with the alphabetical designator. For example, instead of CR15, T2, and F3, the reference designators 15CR, 2T, and 3F are used.

The components in our diagram example shown in Figure 1–18 are numbered with reference designators. The transformer is T1 and the fuse is F1. Other components will be assigned reference designators as they are added to the diagram.

Boolean Logic and Relay Logic

Since the relays in a machine perform some type of control operation, it can be said that they perform a logical function. As with all logical functions, these control circuits must consist of the fundamental AND, OR, and INVERT logical operations. Relay coils, N/C contacts, and N/O contacts can be wired to perform these same fundamental logical functions. By properly wiring relay contacts and coils together, we can create any logical function desired.

AND

Generally, when introducing a class to logical operations, an instructor illustrates the **AND** function using the analogy of a series connection of two switches—a lamp and a voltage source. Relay logic allows the function to be represented this way. Figure 1–20 shows the actual wiring connection for two switches—a lamp and a voltage source in an AND configuration.

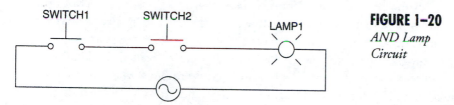

FIGURE 1–20

AND Lamp Circuit

The lamp, LAMP1, will illuminate only when SWITCH1 AND SWITCH2 are ON. The Boolean expression for this is shown in Equation 1–1.

$$Lamp1 = (Switch1) \cdot (Switch2) \qquad (1\text{–}1)$$

To represent the circuit of Figure 1–20 in ladder logic form in an electrical machine diagram, we would utilize the power from the rails and simply add the two switches (we have assumed these are to be pushbutton switches) and lamp in series between the rails as shown in Figure 1–21. This added circuit forms what is called a **rung.** It is called a rung because as we add more circuitry to the diagram, it will begin to resemble a ladder with two uprights and many rungs.

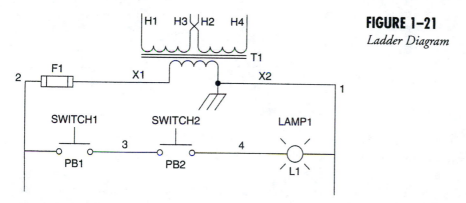

FIGURE 1–21

Ladder Diagram

There are a few important details that have been added along with the switches and lamp. Note that the added wires have been assigned the wire numbers 3 and 4 and the added components have been assigned the reference designators PB1, PB2, and L1. Also note that the switches are on the left, and the lamp is on the right. This is a standard convention when designing and drawing machine circuits. The controlling devices (in this case the switches) are always positioned on the left side of the rung, and the controlled devices (in this case the lamp) are always positioned on the right side of the rung. This wiring scheme is also done for safety reasons. Assume, for example, that we put the lamp on the left side and the switches on the right. Should there develop a short to ground in the wire from the lamp to the switches, the lamp would light without either of the switches being pressed. For a lamp to inadvertently light is not a serious problem, but assume that instead of a lamp we had the coil of a relay that started the machine. This would mean that a short circuit would start the machine without any warning. By properly wiring the controlled device (called the **load**) on the right side, a short in the circuit will cause the fuse to blow when the rung is activated, thus de-energizing the machine controls and shutting down the machine.

OR

A similar approach may be taken for the **OR** function. The circuit shown in Figure 1–22 illustrates two switches wired as an OR function controlling a lamp—LAMP2. As can be seen from the circuit diagram, the lamp will illuminate if either SWITCH1 OR SWITCH2 is closed; that is, depressing either of the switches will cause LAMP2 to illuminate. The Boolean expression for this circuit is illustrated in Equation 1–2.

$$Lamp2 = (Switch1) + (Switch2) \qquad \textbf{(1–2)}$$

FIGURE 1–22

OR Lamp Circuit

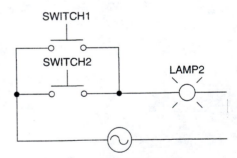

We can now add this circuit to our ladder diagram as another rung, shown in Figure 1–23. Note that since the switches SWITCH1 and SWITCH2 are the same ones used in the top rung, they will have the same names and the same reference designators when drawn in rung 2. This means that each of these two switches has two N/O contacts on the switch assembly. Some designers prefer to place dashed lines between the two PB1 switches and another between the two PB2 switches to clarify that they are operated by the same switch actuator (in this case the actuator is a pushbutton).

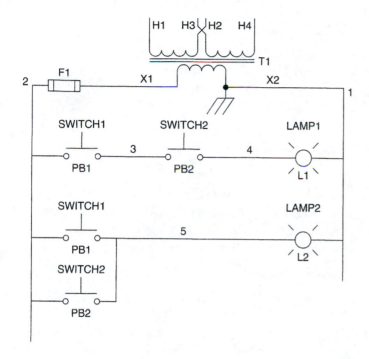

FIGURE 1–23

*Ladder Diagram
with Rung 2
Added*

When we have two or more components in parallel in a rung, each parallel path is called a **branch.** In our diagram in Figure 1–23, rung 2 has two branches, one with PB1 and the other with PB2. It is possible to have branches on the load side of the rung also. For example, we could place another lamp in parallel with LAMP2, thereby creating a branch on the load side.

It is important to note that in our ladder diagram it is possible to exchange rungs 1 and 2 without changing the way the lamps operate. This is one advantage of using ladder diagramming. The rungs can be arranged in any order without changing the way the machine operates. It allows the designer to compartmentalize and organize the control circuitry so that it is easier to understand and troubleshoot. However, keep in mind that, later in this text when we begin PLC ladder programming, the rearranging of rungs is not recommended. In a PLC, the ordering of the rungs is critical, and rearranging the order could change the way the PLC program executes.

AND OR and OR AND

Let us now complicate the circuitry somewhat. Suppose that we add two more switches to the previous circuits and configure the original switch, battery, and light circuit as in Figure 1–24.

FIGURE 1–24

AND-OR Lamp Circuit

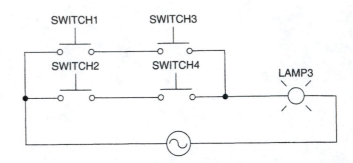

Notice that two switches have been added, SWITCH3 and SWITCH4. In operation, LAMP3 will light if SWITCH1 AND SWITCH3 are both ON, OR if SWITCH2 AND SWITCH4 are both ON. This circuit is called an AND-OR circuit. The Boolean expression for this is illustrated in Equation 1–3.

$$Lamp3 = (Switch1 \cdot Switch3) + (Switch2 \cdot Switch4) \qquad \textbf{(1–3)}$$

The opposite of this circuit, called the OR-AND circuit, is shown in Figure 1–25. For this circuit, LAMP4 will be ON whenever SWITCH1 OR SWITCH2, AND SWITCH3 OR SWITCH4 are ON. For circuits that are logically complicated, it sometimes helps to list all the possible combinations of inputs (switches) that will energize a rung. For this OR-AND circuit, LAMP4 will be lit when the following combinations of switches are ON:

SWITCH1 and SWITCH3

SWITCH1 and SWITCH4

SWITCH2 and SWITCH3

SWITCH2 and SWITCH4

SWITCH1 and SWITCH2 and SWITCH3

SWITCH1 and SWITCH2 and SWITCH4

SWITCH1 and SWITCH3 and SWITCH4

SWITCH2 and SWITCH3 and SWITCH4

SWITCH1 and SWITCH2 and SWITCH3 and SWITCH4

The Boolean expression for the OR-AND circuit is shown in Equation 1–4

$$Lamp3 = (Switch1 + Switch2) \cdot (Switch3 + Switch4) \qquad \textbf{(1–4)}$$

These two rungs will now be added to our ladder diagram and are shown in Figure 1–26. Look closely at the circuit, and follow the possible power paths

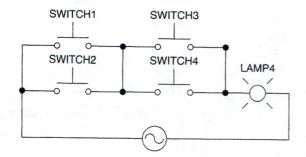

FIGURE 1–25
OR-AND Lamp Circuit

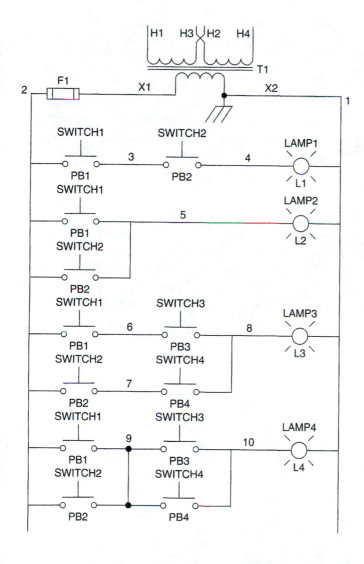

FIGURE 1–26
Ladder Diagram with Rungs 3 and 4 Added

to energize LAMP3 and LAMP4. You should see two possible paths for LAMP3:

SWITCH1 AND SWITCH3

SWITCH2 AND SWITCH4

Either of these paths will allow LAMP3 to energize. For LAMP4, you should see four possible paths:

SWITCH1 AND SWITCH3

SWITCH1 AND SWITCH4

SWITCH2 AND SWITCH3

SWITCH2 AND SWITCH4

Any one of these four paths will energize LAMP4.

Now that we have completed a fundamental study of the ladder diagram, we should begin investigating some standard ladder logic circuits that are commonly used on electrical machinery. Keep in mind that these circuits are also used in programming programmable logic controllers.

Ground Test

Earlier, we drew a ladder diagram of some switch circuits that included the control transformer. We connected the right side of the transformer to ground (the frame of the machine). For safety reasons, it is necessary to occasionally test this ground to be sure that is it still connected because loss of the ground circuit will not affect the performance of the machine and will therefore go unnoticed. This test is done using a ground test circuit and is shown in Figure 1–27.

FIGURE 1–27
Ground Test
Circuit

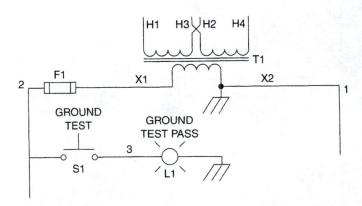

Notice that this rung is unusual in that it does not connect to the right rail. In this case, the right side of lamp L1 has a wire with a lug that is fastened to the frame of the machine under a screw. When the pushbutton S1 is pressed, the lamp L1 lights if there is a path for current to flow through the frame of the machine back to the X2 side of the control transformer. If the lamp fails to light, it is likely that the transformer is no longer grounded. The machine should not be operated until an electrician checks and repairs the problem. In some cases, the lamp L1 is located inside the pushbutton switch S1 (this is called an **illuminated switch**).

The Latch (with Sealing/Latching Contacts)

Occasionally, it is necessary to have a relay "latch" ON so that if the device that activated the relay is switched OFF, the relay remains ON. This is particularly useful for making a momentary pushbutton switch perform as if it were a maintained switch. Consider, for example, the pushbuttons that switch a machine on and off. This can be done with momentary pushbuttons if we include a relay in the circuit that is wired as a latch as shown in the ladder diagram segment of Figure 1–28 (for clarity, the transformer and fuse are not shown). Follow in the diagram as we discuss how this circuit operates.

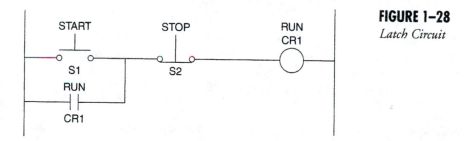

FIGURE 1–28

Latch Circuit

First, when power is applied to the rails, CR1 is initially de-energized and the N/O CR1 contact in parallel with switch S1 is also open. Since we are assuming S1 has not been pressed, there is no path for current to flow through the rung and it will be OFF. Next, we press the START switch S1. This provides a path for current flow through S1, S2, and the coil of CR1, which energizes CR1. As soon as CR1 energizes, the N/O CR1 contact in parallel with S1 closes (since the CR1 contact is operated by the CR1 coil). When the relay contact closes, we no longer need switch S1 to maintain a path for current flow through the rung because it is provided by the N/O CR1 contact and N/C pushbutton S2. At this point, we can release S1 and the relay CR1 will remain energized. The N/O CR1 contact "seals" or "latches" the circuit ON. This type of contact configuration has several commonly used names including a **sealing contact, seal-in contact,** or **latching contact.**

The circuit is de-energized by pressing the STOP switch S2. This interrupts the flow of current through the rung, de-energizes the CR1 coil, and opens the CR1 contact in parallel with S1. When S2 is released, there will still be no current flow through the rung because both S1 and the CR1 N/O contact are open.

The latch circuit has one other feature that cannot be obtained by using a maintained switch. Should power fail while the machine is on, the latch rung will, of course, de-energize. However, when power is restored, the machine will not automatically restart. It must be manually restarted by pressing S1. This is a safety feature that is required on all heavy machines.

Two-Handed Anti-Tie Down, Anti-Repeat

Many machines used in manufacturing are designed to go through a repeated fixed cycle. An example of this is a metal cutter that slices sheets of metal when actuated by an operator. By code, all cyclic machines must have **two-handed RUN** actuation and **anti-repeat** and **anti-tie down** features. Each of these is explained in the following paragraphs.

Two-Handed RUN Actuation   "Two-Handed RUN" means that the machine can only be cycled by an operator pressing two switches simultaneously that are separated by a distance such that both switches cannot be pressed by one hand. This assures that both of the operator's hands will be on the switches and not in the machine when it is cycling. This is simply two palm switches in series operating a RUN relay CR1 as shown in Figure 1–29.

FIGURE 1–29
Two-Handed Operation

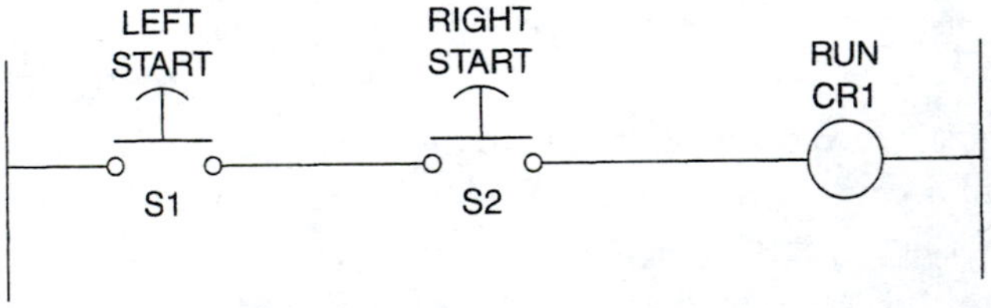

Anti-Tie Down and Anti-Repeat   The machine must not have the capability to be cycled by tying down or taping down one of the two RUN switches and using the second to operate the machine. In some cases, machine operators have done this so that they have one hand available to guide raw material into the machine while it is cycling, which is obviously an extremely hazardous practice. Anti-tie down and anti-repeat go hand-in-hand by forcing both RUN switches to be cycled OFF and then ON each time to make the machine perform one cycle.

This means that both RUN switches must be pressed at the same time within a small time window, usually half a second. If one switch is pressed and then the other is pressed after the time window has expired, the machine will not cycle.

Since both switches must be pressed within a time window, we will need a time delay relay for this feature, specifically a delay-on, or TON, relay. Consider the circuit shown in Figure 1–30. Notice that we have taken the two-handed circuit constructed in Figure 1–29 and added additional circuitry to perform the anti-tie down. Follow along in the circuit diagram as we analyze how it operates.

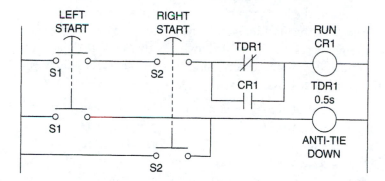

FIGURE 1–30

Two-Handed Operation with Anti-Tie Down and Anti-Repeat

The two palm switches, S1 and S2, now each have two N/O contacts. In the first rung they are connected in series, and in the second rung they are connected in parallel. This means that in order to energize CR1 *both* S1 *and* S2 must be pressed, and in order to energize TDR1 *either* S1 *or* S2 must be pressed. When power is applied to the rails, assuming neither S1 nor S2 are pressed, both relays CR1 and TDR1 will be de-energized. Now, we press either of the two palm switches. Since we did not yet press both switches, relay CR1 will not energize. However, in the second rung, since one of the two switches is pressed, we have a current path through the actuated switch to the coil of TDR1. The time delay relay TDR1 begins to count time. As long as we continue to hold either switch depressed, TDR1 will time-out in half a second. When this happens, the N/C TDR1 contact in the first rung will open, and the rung will be disabled from energizing, which, in turn, prevents the machine from running. At this point, the only way the first rung can be enabled is to first reset the time delay relay by releasing both S1 and S2.

If S1 and S2 are both pressed within half a second of each other, the TDR1 N/C contact in the first rung will have not yet opened and CR1 will be energized. When this happens, the N/O CR1 contact in the first rung seals across the TDR1 contact so that when the time delay relay TDR1 times-out, the first rung will not be disabled. As long as we hold both palm switches ON, CR1 will remain on and TDR1 will remain timed-out.

If we momentarily release either of the palm switches, CR1 de-energizes. When this happens, we lose the sealing contact across the N/C TDR1 contact in the first rung. If we re-press the palm switch, CR1 will not re-energize because TDR1 is still timed-out and is holding its N/C contact open in the first rung. The only way to re-energize CR1 is to reset TDR1 by releasing both S1 and S2 and then pressing both again.

Single Cycle

When actuated, the machine must perform only one cycle and then stop, even if the operator is still depressing the RUN switches. This prevents surprises and possible injury for the operator if the machine should inadvertently go through a second cycle. Therefore, circuitry is usually needed to assure that once the machine has completed one cycle of operation, it stops and waits for the RUN switch(es) to be released and then pressed again.

In order for the circuitry to be able to determine where the machine is in its cycle, a cam-operated limit switch (like the one illustrated in Figure 1–11) must be installed on the machine as shown in Figure 1–31. The cam is mounted on the mechanical shaft of the machine which rotates one revolution for each cycle of the machine. There is a spring inside the switch that pushes the actuator button, lever arm, and roller to the right and keeps the roller constantly pressed against the cam surface. The mechanism is adjusted so that when the cam rotates, the roller of the switch assembly rolls out of the detent in the cam which causes the lever arm to press the switch's actuator button. The actuator remains pressed until the cam makes one complete revolution and the detent again aligns with the roller.

FIGURE 1–31
Cam-Operated Limit Switch

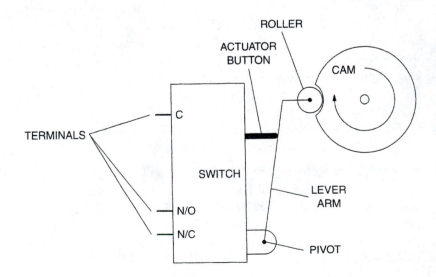

The cam is aligned on the shaft so that when the machine is at the stopping point in its cycle (i.e., between cycles), the switch roller is in the cam detent. The switch has three terminals, C (common, or wiper), N/O (normally open), and N/C (normally closed). When the machine is between cycles, the N/O terminal is open and the N/C is connected to C. While the machine is cycling, the N/O is connected to C and the N/C is open.

The circuit to implement the single-cycle feature is shown in Figure 1–32. Note that we will be using both the N/O and N/C contacts of the cam-operated limit switch LS1. Also note that, for the time being, the START switch S1 is shown as a single pushbutton switch. Later, we will add the two-handed anti-tie down and anti-repeat circuitry to make a complete cycle control system. Follow along on the ladder diagram as we analyze how this circuit works.

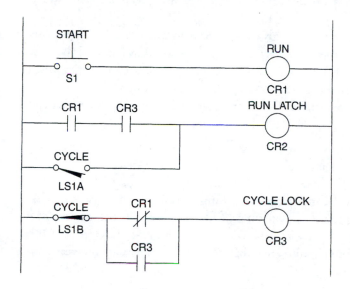

FIGURE 1–32

Single-Cycle Circuit

When the rails are energized, we will assume that the machine is mechanically positioned so that the cam switch is sitting in the cam detent (i.e., the N/O contact LS1A is open and the N/C contact LS1B is closed). At this point, CR1 in the first rung will be OFF (because the START switch has not yet been pressed), CR2 in the second rung is OFF (because CR1 is OFF and LS1A is open), and CR3 in the third rung is ON because LS1B is closed and the N/C CR1 contact is closed. As soon as CR3 energizes, the CR3 N/O contact in the third rung closes. At this point, the circuit is powered and the machine is stopped but ready to cycle.

We now press the START switch S1. This energizes CR1. In the second rung, the N/O CR1 contact closes. Since the N/O CR3 contact is already closed (because

CR3 is ON), CR2 energizes. This applies power to the machine and causes the cycle to begin.

As soon as the cam switch rides out of the cam detent, LS1A closes and LS1B opens. When this happens, LS1A in the second rung seals CR2 ON. In the third rung, LS1B opens, which de-energizes CR3. Since CR2 is still ON, the machine continues in its cycle. The operator may or may not release the START switch during the cycle. However, in either case, it will not affect the operation of the machine. We will analyze both cases.

1. If the operator *does* release the START switch before the machine finishes its cycle, CR1 will de-energize. However, in the second rung this has no immediate effect because the contacts CR1 and CR3 are sealed by LS1A. Also, in the third rung, it has no immediate effect because LS1B is open, which disables the entire rung. Eventually, the machine finishes its cycle and the cam switch rides into the cam detent. This causes LS1A to open and LS1B to close. In the third rung, since the N/C CR1 contact is closed (CR1 is OFF because S1 is released), closing LS1B switches ON CR3. In the second rung, when LS1A opens, CR2 de-energizes (because the N/O CR1 contact is open). This stops the machine and prevents it from beginning another cycle. The circuit is now back in its original state and ready for another cycle.

2. If the operator *does not* release the START switch before the machine finishes its cycle, CR1 remains energized. Eventually, the machine finishes its cycle and the cam switch rides into the cam detent. This causes LS1A to open and LS1B to close. In the third rung, the closing of LS1B has no effect because N/C CR1 is open. In rung 2, the opening of LS1A causes CR2 to de-energize, stopping the machine. Then, when the operator releases S1, CR1 turns OFF, and CR3 turns ON. The circuit is now back in its original state and ready for another cycle.

There are some speed limitations to this circuit. First, if the machine cycles so quickly that the cam switch "flies" over the detent in the cam, the machine will cycle endlessly. One possible fix for this problem is to increase the width of the detent in the cam. However, if this fails to solve the problem, a non-mechanical switch mechanism must be used. Normally, the mechanical switch is replaced by an optical interrupter switch, and the cam is replaced with a slotted disk. This will be discussed in a later chapter. Secondly, if the machine has high inertia, it is possible that it may "coast" through the stop position. In this case, some type of electrically actuated braking system must be added that will quickly stop the machine. For our circuit, the brakes could be actuated by a N/C contact on CR2.

Combined Circuit

Figure 1–33 shows a single-cycle circuit with the START switch replaced by the two rungs that perform the two-handed, anti-tie down, and anti-repeat functions. In this circuit, when both palm switches are pressed within half a second of each other, the machine will cycle once and stop, even if both palm switches remain pressed. Afterward, both palm switches must be released and pressed again in order to make the machine cycle again.

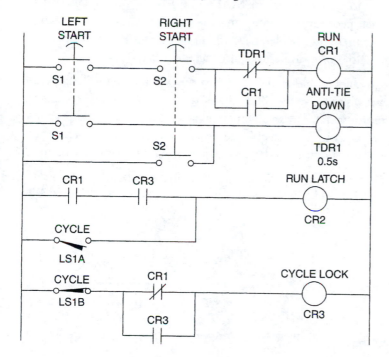

FIGURE 1–33

Two-Handed, Anti-Tie Down, Anti-Repeat, Single-Cycle Circuit

Master Control Relays (MCRs) and Control Zones

Occasionally, the machine controls designer encounters instances where it is convenient to enable or disable entire sections of control circuit (rungs). For example, consider a large manufacturing machine that requires periodic maintenance. When maintenance personnel are working on a machine, it is usually necessary to give them additional capabilities such as the ability to operate the machine with door interlocks defeated, disable annoying alarms, or enable calibration procedures. Obviously, these are unsafe functions that we would not want to be activated during normal operation of the machine. Therefore, one common practice is to provide a **NORMAL / MAINTENANCE selector** switch on the control panel, with a keyed actuator on the switch, so that the MAINTENANCE mode can only be selected by a person who has the key (the maintenance person). The key is captivated in the switch and cannot be removed until it is returned to the NORMAL position.

When these extra functions are included in our control circuit, they must somehow be disabled when the MAINTENANCE switch is OFF. Of course, we could add a N/O contact of the MAINTENANCE switch in series with *every* rung that performs one of these maintenance functions, but that becomes cumbersome to wire, requires many more contacts on the MAINTENANCE switch, and is difficult to modify when necessary. A more convenient way to disable these rungs is to group them together and disable/enable them all as a group with a **master control relay (MCR)** or **zone** (sometimes also called an **interlock**).

The MCR method disables/enables all the remaining rungs in the circuit. This is useful in cases where we wish to give the MCR the capability to shut down the entire remainder of the control circuit. The second method involves the designation of control zones in which we do not wish to disable the remainder of the control circuit, but only a portion of it. It is not unusual to have several zones within a control circuit.

Figure 1–34 illustrates the use of a master control relay in a portion of a control circuit. In this case, START and STOP switches are connected in a latching circuit

FIGURE 1–34

Master Control Relay (MCR) Example

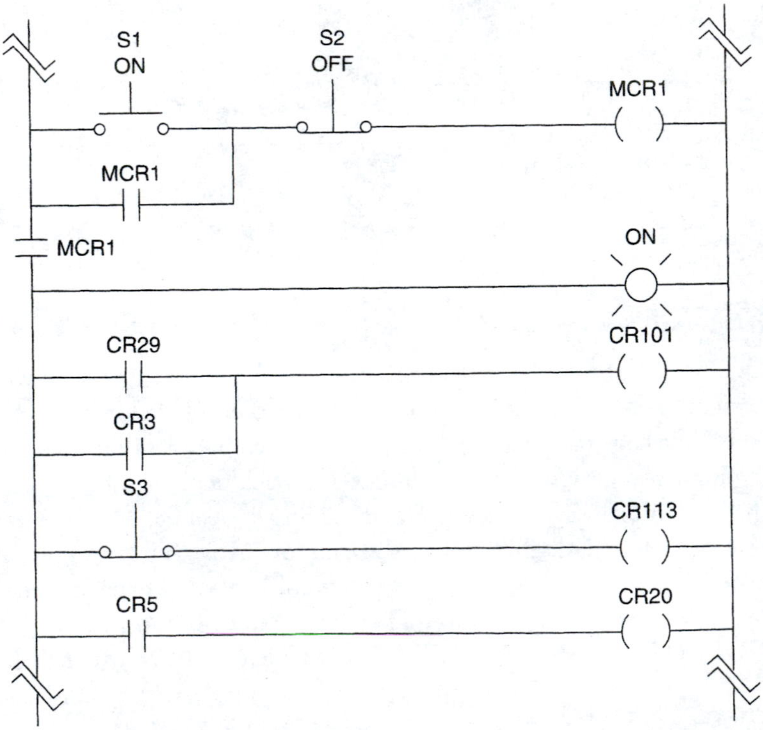

with relay coil MCR1. Then, by using a N/O contact of MCR1 in series with the "hot" rail, we are able to have MCR1 control all the remaining control circuitry.

Zones perform similarly to MCRs except that they are able to apply control to specific sections (or zones) of the control circuitry without directly affecting other areas. Figure 1–35 shows a portion of a control circuit with a zone. Note that the zone is controlled by relay contact CR1 which, in turn, controls a sub-rail. All components connected to the sub-rail are included in the zone and are controlled by the zone control relay. Although the sub-rail in this illustration powers two rungs, the number of rungs allowed on the sub-rail is limited only by the current capacity of the zone control relay contacts. Also, note in our example that the rungs controlling relays CR113 and CR20 are not part of the zone, nor would any rungs added below the control zone.

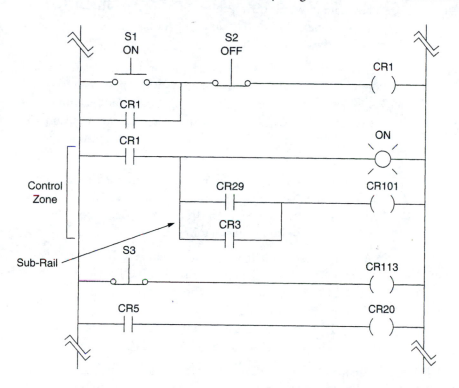

FIGURE 1–35

Control Zone Example

1-3 Machine Control Terminology

There are some words that are used in machine control systems that have special meanings. For safety purposes, the use of these words is explicit and can have no other meaning. They are generally used when naming control circuits, labeling switch positions on control panels, and describing modes of operation of the machine. Following is a list of some of the more important of these terms.

ON   This is a machine state in which power is applied to the machine and to the machine control circuits. The machine is ready to **RUN.** This is also sometimes called the **STANDBY** state.

OFF   Electrically, the opposite of ON. Power is removed from the machine and the machine control circuits. In this condition, pressing any switches on the control panel should have no effect.

RUN   A state in which the machine is cycling or performing the task for which it is designed. This state can only be started by pressing RUN switches. Do not confuse this state with the ON state. It is possible for a machine to be ON but not RUNNING.

STOP   The state in which the machine is ON but not RUNNING. If the machine is RUNNING, pressing the STOP switch will cause RUNNING to cease.

JOG   A condition in which the machine can be "nudged" a small amount to allow for the accurate positioning of raw material while the operator is holding the material. The machine controls must be designed so that the machine cannot automatically go from the JOG condition to the RUN condition while the operator is holding the raw material.

INCH   Similar to JOG, but usually used for conditions that allow the operator to very slowly position the machine in order to improve the accuracy of the operation. INCHing should allow the machine to be positioned but not cycled.

CYCLE   A mode of operation in which the machine RUNs for one complete operation and then automatically STOPs. Holding down the **CYCLE** button will *not* cause the machine to RUN more than one cycle. In order to have the machine execute another CYCLE, the CYCLE button must be released and pressed again. This mode is sometimes called **SINGLE CYCLE.**

TWO-HAND OPERATION   A control design method in which a machine will not RUN or CYCLE unless two separate buttons are simultaneously pressed. This is used on machines where it is dangerous to manually feed raw material into the machine while it is cycling. The two buttons are positioned apart so that they both cannot be pressed by one arm (e.g., a hand and elbow). Both buttons must be released and pressed again to have the machine start another cycle.

Summary

Although this chapter gives the reader a basic understanding of conventional machine controls, it is not intended to be a comprehensive coverage of the subject. Expertise in the area of machine controls can best be achieved by actually practicing the trade under the guidance of experienced machine controls designers. However, an understanding of basic machine controls is the foundation needed to learn the programming language of Programmable Logic Controllers. As we will see in subsequent chapters, the programming language for PLCs is a graphic language that looks very much like machine control electrical diagrams.

Review Questions

1. What is the purpose of the control transformer in machine control systems?

2. Why are fuses necessary in control circuits even though the power mains may already have circuit breakers?

3. What is the purpose of the a) shrouded pushbutton actuator, b) flush pushbutton actuator, and the c) extended pushbutton actuator?

4. Draw the electrical symbol for a two-position selector switch with one contact. The switch is named "ICE" and the selector positions are "CUBES" on the left and "CRUSHED" on the right. The contact is to be closed when the switch is in the "CUBES" position.

5. Draw an electrical diagram rung showing a N/O contact CR5 in series with a N/C contact CR11, operating a lamp L3.

6. Draw a ladder diagram that will perform the following operations:

   a. Energize relay coil CR1 only when CR10 is ON AND CR11 is OFF.

   b. Energize relay coil CR1 only when (CR10 is ON AND CR11 is ON) OR CR12 is OFF.

   c. Energize relay coil CR1 only when (CR10 is OFF AND CR11 is OFF) OR (CR12 is ON AND CR13 is ON).

   d. Energize relay coil CR1 only when CR10 and CR11 are both ON OR both OFF.

   e. Energize relay coil CR1 only when either CR10 OR CR11 are ON, but not both.

   f. *De-energize* relay coil CR1 only when CR10 is ON AND CR11 is OFF (Note: CR1 should be energized under all other conditions).

   g. Energize relay coil CR1 only when any two or more of CR10, CR11, and CR12 are ON.

7. A delay-on (TON) relay has a preset of five seconds. If the coil terminals are energized for eight seconds, how long will its contacts be actuated?

8. If a delay-on (TON) relay with a preset of five seconds is energized for three seconds, explain how it reacts.

9. If a delay-off (TOF) relay with a preset of five seconds is energized for one second, explain how the relay reacts.

10. Draw a ladder diagram rung similar to Figure 1–28 that will cause a lamp L5 to illuminate when relay CR1 is ON, CR2 is OFF, and CR3 is OFF.

11. Draw a ladder diagram rung similar to Figure 1–28 that will cause a lamp L7 to be *OFF* when relay CR2 is ON or when CR3 is OFF. L7 should be ON at all other times. (Hint: Make a table showing all the possible combinations of CR2 and CR3 and mark the combinations that cause L7 to be OFF. All those not marked must be the ones when L7 is ON.)

12. Draw a ladder diagram rung similar to Figure 1–28 that will cause relay CR10 to energize when either CR4 and CR5 are ON, or when CR4 is OFF and CR6 is ON. Then, add a second rung that will cause lamp L3 to illuminate four seconds after CR10 energizes.

The Programmable Logic Controller

CHAPTER

2

Objectives

Upon completion of this chapter, you will know:

- the history of the programmable logic controller.

- why the first PLCs were developed and why they were better than the existing control methods.

- the difference between the open frame, shoebox, and modular PLC configurations and the advantages and disadvantages of each.

- the components that make up a typical PLC.

- how programs are stored in a PLC.

- the equipment used to program a PLC.

- the way that a PLC inputs data, outputs data, and executes its program.

- the purpose of the PLC update.

- the order in which a PLC executes a ladder program.

- how to calculate the scan rate of a PLC.

Introduction

This chapter will introduce the programmable logic controller (PLC) with a brief discussion of its history and development and a study of how the PLC executes a program. A physical description of the various configurations of programmable logic controllers and the functions associated with the different components will follow. The chapter will end with a discussion of the unique way that a programmable logic controller obtains input data, processes it, and produces output data, including a short introduction to ladder logic.

It should be noted that in usage a programmable logic controller is generally referred to as a "PLC" or "programmable controller." Although the term "programmable controller" is generally accepted, it is not abbreviated PC because the abbreviation PC is usually used in reference to a personal computer. As we will see in this chapter, a PLC is by no means a personal computer.

2-1 A Brief History

Early machines were controlled by mechanical means using cams, gears, levers, and other basic mechanical devices. As the complexity grew, so did the need for a more sophisticated control system. This system contained wired relay and switch control elements. These elements were wired as required to provide the control logic necessary for the particular type of machine operation. This was acceptable for a machine that never needed to be changed or modified, but as manufacturing techniques improved and plant changeover to new products became more desirable and necessary, a more versatile means of controlling this equipment had to be developed. Hardwired relay and switch logic was cumbersome and time consuming to modify. Wiring had to be removed and replaced to provide for the new control scheme. This modification was difficult and time consuming to design and install and any small "bug" in the design could be a major problem to correct since it also required rewiring of the system. A new means to modify control circuitry was needed. The development and testing ground for this new means was the U.S. auto industry. The time period was the late 1960s and early 1970s, and the result was the programmable logic controller, or PLC. Automotive plants were confronted with a change in manufacturing techniques every time a model changed and, in some cases, for changes on the same model if improvements had to be made during the model year. The PLC provided an easy way to *reprogram* the wiring rather than actually rewiring the control system.

The PLC that was developed during this time was not very easy to program. The language was cumbersome to write, requiring highly trained programmers. These early devices were merely relay replacements and could do very little else. The PLC has at first gradually, and in recent years rapidly, developed into a sophisticated and highly versatile control system component. Units today are capable of performing complex math functions including numerical integration and differentiation and operate at the fast microprocessor speeds now available. Older PLCs were capable of only handling discrete inputs and outputs (i.e., ON-OFF type signals), while today's systems can accept and generate analog voltages and currents as well as a wide range of voltage levels and pulsed signals. PLCs are also designed to be rugged. Unlike their personal computer cousin, they can typically withstand the vibration, shock, elevated temperatures, and electrical noise to which manufacturing equipment is exposed.

As more manufacturers become involved in PLC production and development, and PLC capabilities expand, the programming language is also expanding. This is necessary to allow the programming of these advanced capabilities. Also, manufacturers tend to develop their own versions of ladder logic language (the language used to program PLCs). This complicates learning to program PLCs in general since one language cannot be learned that is applicable to all types. However, as with other computer languages, once the basics of PLC operation and programming in ladder logic are learned, adapting to the various manufacturers' devices is not a complicated process. Most system designers eventually settle on one particular manufacturer that produces a PLC that is personally comfortable to program and has the capabilities suited to his or her area of applications.

2-2 PLC Configurations

Programmable controllers are much like personal computers in that the user can be overwhelmed by the vast array of options and configurations available. Therefore, when it comes to selecting a PLC for an application, experience is the best teacher. As one gains experience with the various options and configurations available, it becomes less confusing to select the unit that will best perform in a particular application.

Basic PLCs are available on a single printed circuit board as shown in Figure 2–1. They are sometimes called **single board PLCs** or **open frame PLCs.** These are totally self-contained (with the exception of a power supply), and, when installed in a system, they are simply mounted inside a controls cabinet on threaded standoffs. Screw terminals on the printed circuit board allow for the connection of the input, output, and power supply wires. These units are generally not expandable, meaning that extra inputs, outputs, and memory cannot be added to the basic unit. However, some of the more sophisticated models can be linked by cable to expansion boards that provide extra input/output. Therefore, with few exceptions, when using this type of PLC, the system designer must take care to specify a unit that has enough inputs, outputs, and programming capability to handle both the present need of the system and any future modifications that may be required. Single board PLCs are very inexpensive (some less than $100), easy to program, small, and consume little power, but, generally speaking, they do not have a large number of inputs and outputs and have a somewhat limited instruction set. They are best suited to small, relatively simple control applications.

PLCs are also available housed in a single case (sometimes referred to as a **shoe box**) with all input and output, power and control connection points located on the single unit as shown in Figure 2–2. These are generally chosen according to available program memory and required number and voltage levels of inputs and outputs to suit the application. These systems generally have an ex-

FIGURE 2–1
*Open Frame PLC
(Courtesy of
Triangle Research
International, Inc.)*

FIGURE 2–2
*Shoebox-Style
PLCs
(Courtesy of IDEC
Corporation)*

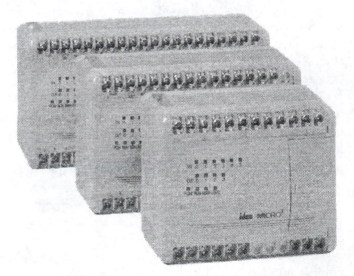

pansion port (an interconnection socket) that will allow the addition of specialized units such as high-speed counters and analog input and output units or additional discrete inputs or outputs. These expansion units are either plugged directly into the main case or connected to it with ribbon cable or other suitable cable.

More sophisticated units, with a wider array of options, are **modularized.** An example of a modularized PLC is shown in Figure 2–3.

The typical system components for a modularized PLC are:

1. Processor. The processor (sometimes called a CPU), as in the self-contained units, is generally specified according to memory required for the program to be implemented. In the modularized versions, capability can also be a factor. This includes features such as higher math functions, PID control loops, and optional programming commands. The processor consists of the microprocessor, system memory, serial communication ports for printer, PLC LAN link and external programming device, and, in some cases, the system power supply to power the processor and I/O modules.

2. Mounting rack. This is usually a metal framework with a printed circuit board backplane that provides the means for mounting the PLC I/O modules and processor. Mounting racks are specified according to the number of modules required to implement the system. The mounting rack provides data and power connections to the processor and modules via the backplane. For CPUs that do not contain a power supply, the rack also holds the modular power supply. There

FIGURE 2–3

Modularized PLC (Courtesy of Omron Electronics, LLC)

are systems in which the processor is mounted separately and connected by cable to the rack. The mounting rack can be available to mount directly to a panel or can be installed in a standard 19-wide equipment cabinet. Mounting racks are cascadable so several may be interconnected to accommodate a large number of I/O modules.

3. Input and output modules. Input and output (I/O) modules are specified according to the input and output signals associated with the particular application. These modules fall into the categories of discrete, analog, high-speed counter, or register types.

Discrete I/O modules are generally capable of handling 8 or 16 and, in some cases 32, ON-OFF type inputs or outputs per module. Modules are specified as input or output but generally not both although some manufacturers now offer modules that can be configured with both input and output points in the same unit. The inputs or outputs of the module can be specified as AC only, DC only, or AC/DC along with the voltage values for which it is designed.

Analog input and output modules are specified according to the desired resolution and voltage or current range. As with discrete modules, these are generally input or output; however, some manufacturers provide analog input and output in the same module. Analog modules that can directly accept thermocouple inputs for temperature measurement and monitoring by the PLC are also available.

Pulsed inputs to the PLC can be accepted using a high-speed counter module. This module can be capable of measuring the frequency of an input signal from a tachometer or other frequency generating device. These modules can also count the incoming pulses if desired. Generally, both frequency and count are available from the same module at the same time if both are required in the application.

Register input and output modules transfer 8 or 16 bit words of information to and from the PLC. These words are generally numbers (BCD or Binary) generated from thumbwheel switches or encoder systems for input or data to be output to a display device by the PLC.

Other types of modules may be available depending upon the manufacturer of the PLC and its capabilities. These include specialized communication modules to allow for the transfer of information from one controller to another. One development allows the serial transfer of information to remote I/O units as far as 12,000 feet away.

4. Power supply. The power supply specified depends upon the manufacturer's PLC being utilized in the application. As stated previously, in some cases, a power supply capable of delivering all required power for the system is furnished as part of the processor module. If the power supply is a separate module, it must be capable of delivering a current greater than the sum of all the currents needed by the other modules. For systems with the power supply inside the CPU module, there may be some modules in the system that require excessive power not available from the processor because of either voltage or current requirements that can only be achieved through the addition of a second power source. This is generally true if analog or external communication modules are present since these require ± DC supplies which, in the case of analog modules, must be well regulated.

5. Programming unit. The programming unit allows the engineer or technician to enter and edit the program to be executed. In its simplest form, it can be a hand-held device with a keypad for program entry and a display device (LED or LCD) for viewing program steps or functions as shown in Figure 2–4. More advanced systems employ a separate personal computer which allows the programmer to write, view, and edit the program and download it to the PLC. This is accomplished with proprietary software available from the PLC manufacturer. This software also allows the programmer or engineer to monitor the PLC as it is running the program. With this monitoring system, such things as internal coils, registers, timers, and other items not visible externally can be monitored to determine proper operation. Also, internal register data can be altered, if required, to fine tune program operation that can be advantageous when debugging the program. Communication with the programmable controller with this system is done via a cable connected to a special programming port on the controller. Connection to the personal computer can be through a serial port or from a dedicated card installed in the computer.

FIGURE 2–4

Programmer Connected to a Shoebox PLC (Courtesy of IDEC Corporation)

2-3 System Block Diagram

A programmable controller is a specialized computer. Since it is a computer, it has all the basic component parts contained in any other computer—a central processing unit, memory, input interfacing, and output interfacing. A typical programmable controller block diagram is shown in Figure 2–5.

FIGURE 2–5

Programmable Logic Controller Block Diagram

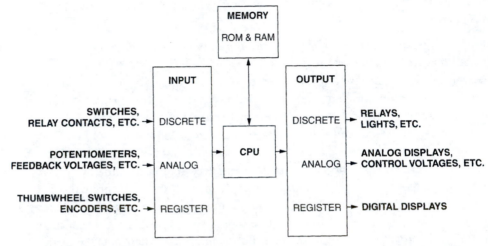

The central processing unit (CPU) is the control portion of the PLC. It interprets the program commands retrieved from memory and acts on those commands. In present day PLCs, this unit is a microprocessor based system. The CPU is housed in the processor module of modularized systems.

Memory in the system generally consists of ROM and RAM. The ROM memory contains the program information that allows the CPU to interpret and act on the ladder logic program stored in the RAM memory. RAM memory is generally kept alive with an on-board battery so that the ladder program is not lost when the system power is removed. This battery can be a standard dry cell or rechargeable nickel-cadmium type. Newer PLC units are now available with Electrically Erasable Programmable Read-Only Memory (EEPROM), which does not require a battery. In modular systems, memory is housed in the processor module.

Input units can be any of several different types depending on input signals expected as described previously. The input section can accept discrete or analog signals of various voltage and current levels. Present day controllers offer discrete signal inputs of both AC and DC voltages from TTL (5 volts) to 250 VDC and from 5 to 250 VAC.

Analog input units can accept input levels such as ±10 VDC, ±5 VDC and 4–20 ma. current loop values. Discrete input units present each input to the CPU as a single 1 or 0 while analog input units contain analog to digital conversion circuitry and present the input voltage to the CPU as a binary number normalized to the maximum count available from the unit. The number of bits representing the input voltage or current depends upon the resolution of the unit. This number generally contains a defined number of magnitude bits and a sign bit. Register input units present the word input to the CPU as it is received (Binary or BCD).

Output units operate much the same as the input units with the exception that the unit is either sinking (supplying a ground) or sourcing (providing a voltage) discrete voltages or sourcing analog voltage or current. These output signals are presented as directed by the CPU. The output circuit of discrete units can be transistors for TTL and higher DC voltage or thyristors for AC voltage outputs. For higher current applications and situations where a physical contact closure is required, mechanical relay contacts are available. These higher currents, however, are generally limited to about 2–3 amperes. The analog output units have internal circuitry that performs the digital to analog conversion and generates the variable voltage or current output.

2-4 Update - Solve the Ladder - Update

When power is applied to a programmable logic controller, the PLC's operation consists of two steps: (1) update inputs and outputs and (2) solve the ladder. This may seem like a very simplistic approach to something that has to be more complicated, but truly there are only these two steps. If these two steps are thoroughly understood, writing and modifying programs and getting the most from the device is much easier to accomplish. With this understanding, the things that can be undertaken are then up to the imagination of the programmer.

The "update - solve the ladder" sequence begins after startup. The actual startup sequence includes some operations transparent to the user or programmer that occur before actual PLC execution of the user program begins. During this startup there may be extensive diagnostic checks performed by the processor on things such as memory, I/O devices, communication with other devices (if present), and program integrity. In sophisticated modular systems, the processor is able to identify the various module types, their location in the system, and address. This type of system analysis and testing generally occurs during startup before actual program execution begins.

2-5 Update

The first thing the PLC does when it begins to function is update I/O. This means that all discrete input states are recorded from the input unit and all discrete states to be output are transferred to the output unit. Register data generally has specific addresses associated with it for both input and output data, referred to as input and output registers. These registers are available to the input and output modules requiring them and are updated with the discrete data. Since this is input/output updating, it is referred to as **I/O update.** The updating of discrete input and output information is accomplished with the use of input and output image registers set aside in the PLC memory. Each discrete input point has associated with it one bit of an input image register. Likewise, each discrete output point has one bit of an output image register associated with it. When I/O updating occurs, each input point that is ON at that time will cause a 1 to be set in the image register at the bit address associated with that particular input. If the input is OFF, a 0 will be set into the bit address. Memory in today's PLCs is generally configured in 16-bit words. This means that one word of memory can store the states of 16 discrete input points. Therefore, there may be a number of words of memory set aside as the input and output image registers. At I/O update, the status of the input image register is set according to the state of all discrete inputs and the status of the output image register is transferred to the output unit. This transfer of information typically only occurs at I/O update. It may be forced to occur at other times in PLCs that have an Immediate I/O Update command. This command will force the PLC to update the I/O at other times, although this would be a special case.

One major item of concern about the first output update is the initial state of outputs. This is a concern because there may be outputs that, if initially turned ON, could create a safety hazard, particularly in a system controlling heavy mechanical devices capable of injuring personnel. In some systems, all outputs may need to be initially set to their OFF state to ensure the safety of the system. However, there may be systems that require outputs to initially be setup in a specific way—some ON and some OFF. This could take the form of a predetermined setup or could be a requirement that the outputs remain in the state immediately before power-down. More recent systems have provisions for both setup options and even a combination of the two. This is a prime concern of the engineer and programmer and must be defined as the system is being developed to ensure the safety of personnel that operate and maintain the equipment. Safety as related to system and program development will be discussed in a later chapter.

2-6 Solve the Ladder

After the I/O update has been accomplished, the PLC begins executing the commands in the user program. These commands are typically referred to as the ladder diagram. The ladder diagram is basically a representation of the program steps using relay contacts and coils. The ladder is drawn with contacts to the left side of the page (or screen) and coils to the right. This is a holdover from the time when control systems were relay based, and this type of diagram was used for the electrical schematic of those systems. A sample ladder diagram (or ladder program) is shown in Figure 2–6.

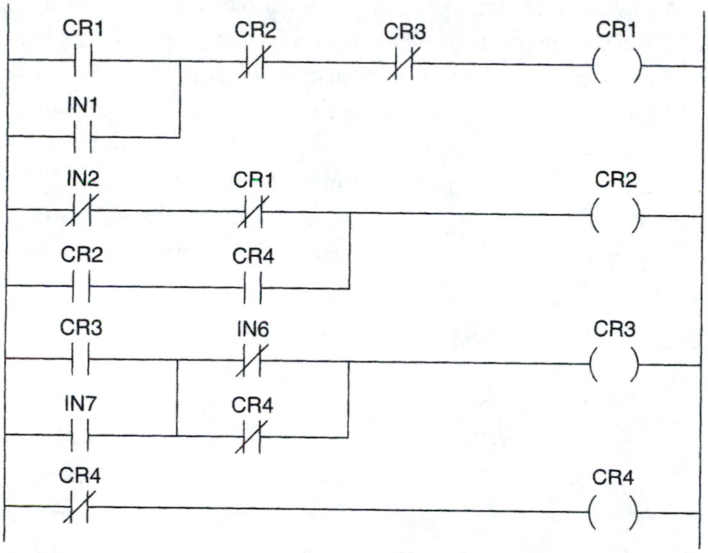

FIGURE 2–6
Sample PLC Ladder Diagram

The symbols used in Figure 2–6 may be foreign at this point so a short explanation will be necessary. The symbols at the right of the ladder diagram, which are circular and labeled CR1, CR2, CR3, and CR4, are the software coils of the relays. The symbols at the left that look like capacitors, some with diagonal lines through them, are the contacts associated with the coils. The symbols that look like capacitors without the diagonal lines through them are normally open contacts. These are analogous to a switch that is normally OFF. When the switch is turned ON, the contact closes. The contact symbols at the left that look like capacitors with diagonal lines through them are normally closed contacts. A normally closed contact is equivalent to a switch that is normally turned ON. It will turn OFF when the switch is actuated.

As can be seen in Figure 2–6, contact and coil position are as described previously. Also, one can see the reason for the term ladder diagram if the rungs of a stepladder are visualized. In fact, each complete line of the diagram is referred to as one rung of logic. The actual interpretation of the diagram will be discussed later although some explanation is required here. The contact configuration on the left side of each rung can be visualized as switches and the coils on the right as lights. If the switches are turned ON and OFF in the proper configuration, the light to the right will illuminate. The PLC executes this program from left to right and top to bottom, in that order. It first looks at the switch (contact) configuration to determine if current can be passed to the light (coil). The data for this decision comes from the output and input image registers. If current can be passed, the light (coil) will then be turned ON. If not, the light (coil) will be turned OFF. This is recorded in the output image register. Once the PLC has analyzed at the left side of the rung, it ignores the left side of the rung until the next time it solves that particular rung. Once the light (coil) has been either turned ON or OFF, it will remain in that state until the next time the PLC solves that particular rung. After solving a rung, the PLC moves on to solve the next rung in the same manner, and the process continues until the entire ladder has been executed and solved. One rule that is different from general electrical operation is the direction of current flow in the rung. In a ladder logic rung, current can only flow from left to right and up and down; never from right to left.

As an example, in the ladder shown in Figure 2–7, coil CR1 will energize if any of the following conditions exist:

FIGURE 2–7

Illustration of Allowed Current Flow in a Ladder Rung

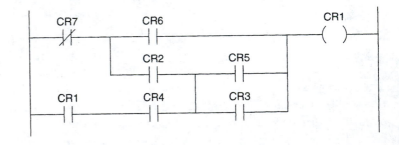

1. CR7 is OFF, CR6 is ON.

2. CR7 is OFF, CR2 is ON, CR5 is ON.

3. CR7 is OFF, CR2 is ON, CR3 is ON.

4. CR1 is ON, CR4 is ON, CR3 is ON.

5. CR1 is ON, CR4 is ON, CR5 is ON.

You will notice that the current flow in the circuit in each of the cases listed is from left to right and up and down. CR1 will not energize in the following case:

CR1 is ON, CR4 is ON, CR2 is ON, CR6 is ON, CR5 is OFF, CR3 is OFF, CR7 is ON.

This is because current would have to flow from right to left through the CR2 contact. This is not allowed in ladder logic even though current could flow in this direction if we were to build it with real relays. Remember, we are working in the software world, not the hardware world.

To review, after the I/O update, the PLC moves to the first rung of ladder logic. It solves the contact configuration to determine if the coil is to be energized or de-energized. It then energizes or de-energizes the coil. After this is accomplished, it moves to the left side of the next rung and repeats the procedure. This continues until all rungs have been solved. When this procedure is complete, with all rungs solved and all coils in the ladder set up according to the solution of each rung, the PLC proceeds to the next step of its sequence, which is the I/O update.

At I/O update, the state of all coils that are designated as outputs are transferred from the output image register to the output unit and the state of all inputs are transferred to the input image register. Note that any input changes that occur during the solution of the ladder are ignored because they are only recorded at I/O update time. The state of each coil is recorded to the output image register as each rung is solved. *However, these states are not transferred to the output unit until I/O update time.*

This procedure of I/O update and solving the ladder diagram is referred to as scanning and is represented in Figure 2–8. The period between one I/O update and the next is referred to as one **scan.** The amount of time it takes the PLC to get from one I/O update to the next is referred to as **scan time.** Scan time is typically measured in milliseconds and is related to the speed of the CPU, the length of the ladder diagram, and the types of instructions in the program. The slower the processor or the longer the ladder diagram, the longer the scan time of the system. The speed at which a PLC scans is referred to as **scan rate.** Scan rate units are usually listed in msec/K of memory being utilized for the program. As an example, if a particular PLC has a scan rate of 8 msec/K and the program occupies 6K of memory, it will take the PLC 48 msec to complete one scan of the program.

FIGURE 2–8
Scan Cycle

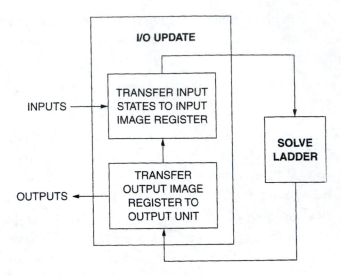

Summary

Before a study of PLC programming can begin, it is important to gain a fundamental understanding of the various types of PLCs available, the advantages and disadvantages of each, and the way in which a PLC executes a program. The open frame, shoebox, and modular PLCs are each best suited to specific types of applications based on the environmental conditions, number of inputs and outputs, ease of expansion, and method of entering and monitoring the program. Additionally, programming requires a prior knowledge of the manner in which a PLC receives input information, executes a program, and sends output information. With this information, we are now prepared to begin a study of PLC programming techniques.

Review Questions

1. How were early machines controlled before PLCs were developed?

2. When were the first PLCs developed?

3. What is an open frame PLC? A shoebox PLC? A modular PLC?

4. List four types of I/O modules.

5. List five devices that would be typical inputs to a PLC. List five devices that a PLC might control.

6. What types of memory might a PLC contain?

7. Which type or types of memory would store the program to be executed by the PLC?

8. What is the purpose of the programming unit?

9. What type of control system did the PLC replace? Why was the PLC better?

10. What industry was primarily responsible for PLC development?

11. What are the two steps the PLC must perform during operation?

12. Describe I/O update.

13. What is the Output Image Register?

14. Describe the procedure for solving a rung of logic.

15. What is the allowed direction of current flow in a ladder logic rung?

16. Define scan rate.

17. If a PLC program is 7.5K long and the scan rate of the machine is 7.5 msec/K, what will be the length of time between I/O updates?

18. Define scan time.

19. At what time is data transferred to and from the outside world into a PLC system?

20. What common devices may be used to understand the operation of coils and contacts in ladder logic?

Fundamental PLC Programming

Objectives

Upon completion of this chapter, you will know:

- how to convert a simple electrical ladder diagram to a PLC program.

- the difference between physical components and program components.

- why using a PLC saves on the number of physical components.

- how to construct disagreement and majority circuits.

- how to construct special purpose logic in a PLC program, such as the oscillator, gated oscillator, sealing contact, and the always-on and always-off contacts.

Introduction

When writing programs for PLCs, it is beneficial to have a background in ladder diagramming for machine controls. This is basically the material that was covered in Chapter 1 of this text. The reason is that, at a fundamental level, ladder logic programs for PLCs are very similar to electrical ladder diagrams. This is not a coincidence. The engineers that developed the PLC programming language were sensitive to the fact that most engineers, technicians, and electricians who work with electrical machines on a day-to-day basis would be familiar with this method of representing control logic. This would allow someone new to PLCs, but familiar with control diagrams, to be able to very quickly adapt to the programming language. It is likely that PLC programming language is one of the easiest programming languages to learn.

In this chapter, we will take the foundation knowledge learned in Chapter 1 and use it to build an understanding of PLC programming. The programming method used in this chapter will be the graphical method, which uses schematic symbols for relay coils and contacts. In a later chapter we will discuss a second method of programming PLCs, which is the mnemonic language method.

3–1 Physical Components vs. Program Components

When learning PLC programming, one of the most difficult concepts to grasp is the difference between **physical components** and **program components.** We will be connecting physical components (switches, lights, relays, etc.) to the external terminals on a PLC. Then, when we program the PLC, any physical components connected to the PLC will be represented in the program as program components. A program component will not have the same reference designator as the physical component but can have the same name. As an example, consider a N/O pushbutton switch S1 named START. If we connect this to input 001 of a PLC, then when we program the PLC, the START switch will become a N/O relay contact with reference designator IN001 and the name START. As another example, if we connect a RUN lamp L1 to output 003 on the PLC, then in the program, the lamp will be represented by a relay coil with reference designator OUT003 and name RUN (or, if desired, "RUN_LAMP").

As a programming example, consider the simple AND circuit shown in Figure 3–1, which consists of two momentary pushbuttons in series operating a lamp. Although

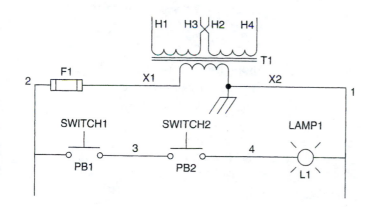

FIGURE 3–1
*AND Ladder
Diagram*

it would be very uneconomical to implement a circuit this simple using a PLC, for this example we will do so.

When we convert a circuit to run on a PLC, we first remove the components from the original circuit and wire them to the PLC as shown in Figure 3–2. One major difference in this circuit is that the two switches are no longer wired in series. Instead, each one is wired to a separate input on the PLC. As we will see later, the two switches will be connected in series in the PLC program. By providing each switch with a separate input to the PLC, we gain the maximum amount of flexibility. In other words, by connecting them to the PLC in this fashion, we can "wire" them in software in any desired fashion.

FIGURE 3–2

PLC Wiring Diagram for Implementation of Figure 3–1

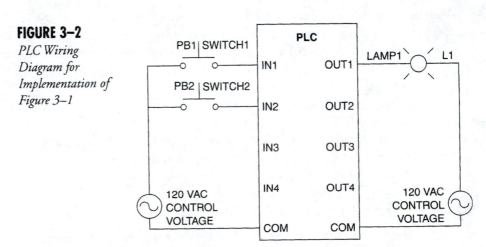

The two 120 V control voltage sources are actually the same source (i.e., the control transformer secondary voltage). They are shown separately in this figure to make it easier to see how the inputs and output are connected to the PLC and how each is powered.

Once we know how the external components are wired to the PLC, we can then write our program. In this case, we need to connect the two switches in series. However, once the signals are inside the PLC, they are assigned new reference designators that are determined by the respective terminal on the PLC to which they are connected. Since SWITCH1 is connected to IN1, it will be called IN1 in our program. Likewise, SWITCH2 will become IN2 in our program. Also, since LAMP1 is connected to OUT1 on the PLC, it will be called relay OUT1 in our program. Our program to control LAMP1 is shown in Figure 3–3.

```
|  IN1       IN2                                                              OUT1
|---| |------| |----------------------------------------------------------(OUT)|
|
|
```

FIGURE 3–3
*AND PLC
Program*

The appearance of the PLC program may look a bit unusual. This is because this ladder rung was drawn by a computer using ASCII characters instead of graphic characters. Notice that the rails are drawn with vertical line characters, the conductors are hyphens, and the coil of OUT1 is made of two parentheses. Also, notice that the right rail is all but missing. Many programs that are used to write and edit PLC ladder programs leave out the rails. The program that produced this diagram leaves out the right rail but puts in the left one with a rung number next to each rung.

When the program shown in Figure 3–3 is run, the PLC first updates the input image register by storing the values of the inputs on terminals IN1 and IN2 (it stores a 1 if an input is ON, and a 0 if it is OFF). Then it solves the ladder diagram according to the way it is drawn and based on the contents of the input image register. For our program, if both IN1 and IN2 are ON, it turns on OUT1 in the output image register. (Careful, it does *NOT* turn on the output terminal yet!) Then, when it has completed solving the entire program, it performs another update. This update transfers the contents of the output image register (the most recent results of solving the ladder program) to the output terminals. This turns on terminal OUT1, which turns on the lamp LAMP1. At the same time that it transfers the contents of the output image register to the output terminals, it also transfers the logical values on the input terminals to the input image register. Now it is ready to solve the ladder again.

For an operation this simple, this is a lot of trouble and expense. However, as we add to our program, we will begin to see how a PLC can economize not only on wiring but on the complexity (and cost) of external components.

Next, we will add another lamp that switches ON when either SWITCH1 or SWITCH2 is ON. If we were to add this circuit to our electrical diagram in Figure 3–1, we would have the circuit shown in Figure 3–4.

Notice that to add this circuit to our existing circuit, we had to add additional contacts to the switches PB1 and PB2. Obviously, this increases the cost of the switches. However, when doing this on a PLC, it is much easier and less costly.

The PLC wiring diagram to implement both the AND and OR circuits is shown in Figure 3–5. Notice here that the only change made to the circuit is to add LAMP2 to the PLC output OUT2. Since the SWITCH1 and SWITCH2 signals

FIGURE 3–4
OR Circuit

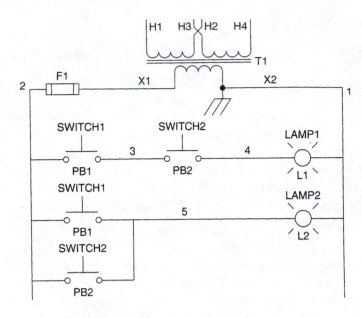

FIGURE 3–5
*PLC Wiring
Diagram to Add
Lamp LAMP2*

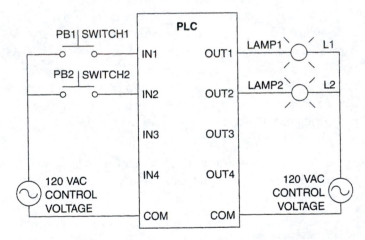

are already available inside the PLC via inputs IN1 and IN2, it is not necessary to bring them into the PLC again. This is because of one unique and very economical feature of PLC programming. *Once an input signal is brought into the PLC for use by the program, you may use as many contacts of the input as you wish, and the contacts may be of either N/O or N/C polarity.* This reduces the cost because, even though our program will require more than one contact of IN1 and IN2, each of the actual switches that generate these inputs, PB1 and PB2, only need to have a single N/O contact.

Now that we have the LAMP2 connected, we can write the program. We do this by simply adding to the program the additional rung needed to perform the OR operation. This is shown in Figure 3–6. Keep in mind that other than the additional lamp and the time it takes to add the additional program, this added OR feature costs nothing.

```
|   IN1       IN2                                                          OUT1
1---| |-------| |---------------------------------------------------------(OUT)|
|
|
|
|   IN1                                                                   OUT2
2---| |------------------------------------------------------------------ (OUT)|
|   IN2   |
|---| |---+
|
|
```

FIGURE 3–6

PLC Program with Added OR Rung

Next, we will add AND-OR and OR-AND functions to our PLC. For this, we will need two more switches, SWITCH3 and SWITCH4, and two more lamps, LAMP3 and LAMP4. Figure 3–7 shows the addition of these switches and lamps.

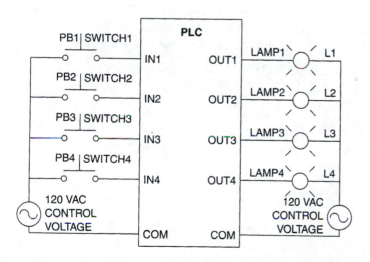

FIGURE 3–7

PLC Wiring Diagram Adding Switches PB3 and PB4, and Lamps LAMP3 and LAMP4

Once the connection scheme for these components is known, we can add the additional programming required to implement the AND-OR and OR-AND functions as shown in Figure 3–8.

FIGURE 3–8
PLC Program with AND-OR and OR-AND Added

```
|  IN1      IN2                                                              OUT1
1---|  |--------|  |-------------------------------------------------------------(OUT)|
|
|
|
|  IN1                                                                       OUT2
2---|  |--------------------------------------------------------------------------(OUT)|
|  IN2   |
|---|  |---+
|
|
|  IN1      IN2                                                              OUT3
3---|  |--------|  |----------------------------------------------------------------(OUT)|
|  IN3      IN4   |
|---|  |--------|  |---+
|
|
|
|  IN1      IN3                                                              OUT4
4---|  |--------|  |----------------------------------------------------------------(OUT)|
|  IN2   |  IN4   |
|---|  |--------|  |---+
|
```

3-2 Example Problem—Lighting Control

A lighting control system is to be developed. The system will be controlled by four switches—SWITCH1, SWITCH2, SWITCH3, and SWITCH4. These switches will control the lighting in a room based on the following criteria:

1. Any of the three switches (SWITCH1, SWITCH2, and SWITCH3), if turned ON, can turn the lighting ON, but all three switches must be OFF before the lighting will turn OFF.

2. The fourth switch, SWITCH4, is a master control switch. If this switch is in the ON position, the lights will be OFF, and none of the other three switches have any control.

Example 3–1
Problem:

Design the wiring diagram for the controller connections, assign the inputs and outputs, and develop the ladder diagram that will accomplish the task.

Solution:

The first task is to draw the controller wiring diagram. To do this, connect all switches to inputs and the lighting to an output, and note the number of inputs and outputs associated with these

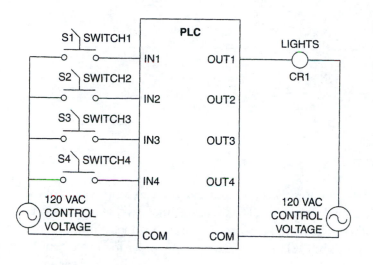

FIGURE 3–9
*PLC Wiring
Diagram for
Example 3–1*

connections. The remainder of the task is in developing the ladder diagram. The wiring diagram is shown in Figure 3–9.

Notice that all four switches are shown as normally open selector switches and the output is connected to a relay coil CR1. We are using the relay CR1 to operate the lights because generally the current required to operate a bank of room lights is higher than the maximum current a PLC output can carry. Attempting to operate the room lights directly from the PLC output will most likely damage the PLC.

For this wiring configuration, the following definition list is apparent:

INPUT IN1 = SWITCH1

INPUT IN2 = SWITCH2

INPUT IN3 = SWITCH3

INPUT IN4 = SWITCH4 (Master Control Switch)

OUTPUT OUT1 = Lights control relay coil CR1

This program requires that when SWITCH4 is ON, the lights must be OFF. In order to do this, it would appear that we need a N/C SWITCH4, not a N/O as we have in our wiring diagram. However, keep in mind that once an input signal is brought into a PLC, we may use as many contacts of the input as we need in our program, *and the contacts may be either N/O or N/C.* Therefore, we may use a N/O switch for SWITCH4, and then in the program we will logically invert it by using N/C IN4 contacts.

The ladder diagram to implement this example problem is shown in Figure 3–10.

First, note that this ladder diagram looks smoother than previous ones. This is because, although it was created using the same program, the ladder was printed using graphics characters (extended ASCII characters) instead of standard ASCII characters.

FIGURE 3–10

*Example 3–1
Lighting Control
Problem*

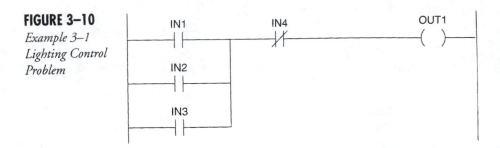

Notice the normally closed contact for IN4. A normally closed contact represents an inversion of the assigned element, in this case IN4, which is defined as SWITCH4. Remember, SWITCH4 has to be in the OFF position before any of the other switches can take control. In the OFF position, SWITCH4 is open. This means that IN4 will be OFF (de-energized). So, in order for an element as signed to IN4 to be closed with the switch in the OFF position, it must be shown as a normally closed contact. When SWITCH4 is turned ON, the input, IN4, will become active (energized). If IN4 is ON, the normally closed IN4 contact will open. With this contact open in the ladder diagram, none of the other switches will be able to control the output. REMEMBER: A normally closed switch will open when energized and close when de-energized.

3-3 Internal Relays

When considering program components, the PLC programmer has additional flexibility available by using **internal relays.** Internal relays are general purpose relays that are internal to the PLC (i.e., they are not directly accessible from the inputs to the PLC, nor can they be used as outputs). As with other program components, they are not "real" relays in a physical sense, but instead are each a digital bit that is stored in an internal image register. From a programming standpoint, all internal relays have one coil and as many N/O and/or N/C contacts as the programmer needs. *All* PLCs have internal relays; however, the internal numbering scheme for them and the maximum allowable quantity depends on the PLC brand and model. Therefore, the programmer must refer to the technical manual for the PLC in order to determine how to reference them in a program and the maximum allowable number that is available. Internal relays are a valuable programming tool. They allow the programmer to perform more complex internal operations without needlessly using costly output relays. In the programming examples in this text, all internal relays are designated with a "CR" prefix followed by a number (e.g., CR1, CR2, etc.).

3-4 Disagreement Circuit

Occasionally, a program rung may be needed that produces an output when two signals disagree (one signal is a logical 1 and the other a logical 0). For example, assume we have two signals, A and B. We would like to produce a third signal, C, under the condition A = 0, B = 1 or A = 1, B = 0. Those familiar with digital logic will recognize this as the **exclusive OR** operation in which the expression is $C = \overline{A}B + A\overline{B} = A \oplus B$. This can also be implemented in ladder logic. Assume the two signals are inputs IN1 and IN2 and the result is OUT1. In this case, the **disagreement circuit** will be as shown in Figure 3–11.

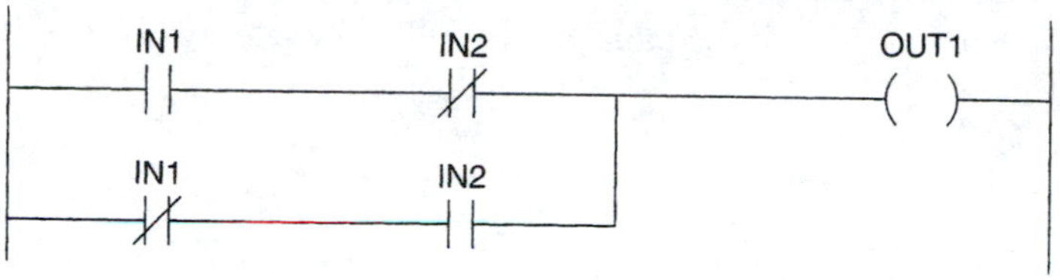

FIGURE 3–11
Disagreement Circuit

For this program, OUT1 will be OFF whenever IN1 and IN2 have the same value (i.e., either both ON or both OFF), and OUT1 will be ON when IN1 and IN2 have different values (i.e., either IN1 ON and IN2 OFF, or IN1 OFF and IN2 ON).

3-5 Majority Circuit

There are situations in which a PLC must make a decision based on the results of a majority of inputs. For example, assume that a PLC is monitoring five tanks of liquid and must give a warning to the operator when three of them are empty. It doesn't matter which three tanks are empty, only that any three of the five are empty. As it turns out, by using binomial coefficients, there are ten possible combinations of three empty tanks. There are also combinations of four empty tanks and the possibility of five empty tanks, but as we will see, those cases will be automatically included when we design the system for three empty tanks.

It is important when designing majority circuits to design them so that "votes" of more than a marginal majority will also be accepted. For example, let's assume that in our five-tank example, the tanks are labeled A, B, C, D, and E, and when an input from a tank is ON, it indicates that the tank is empty. One combination of three empty tanks would be tanks A, B, and C empty and D and E not empty. If this is expressed as a Boolean expression, it would be ABCD'E'. However, this expression would not be true if A, B, C, and D were ON, nor would it be true if all of the inputs were ON. However, leaving D and E as "don't cares" will take into account these possibilities. Therefore, we would shorten our expression to ABC, which would cover the conditions ABCD'E', ABCDE', ABCD'E, and ABCDE; all of which would be majority conditions. It turns out that if we write our program to cover all the conditions of three empty tanks, and each expression uses only three inputs, we will cover (by virtue of "don't cares") the four combinations of four empty tanks and the one combination of five empty tanks.

To find all the possible combinations of three empty tanks out of five, we begin by constructing a binary table of all possible 5-bit numbers, beginning with 00000 and ending with 11111, and we assign each of the five columns to one of the tanks. To the right of the columns, we make another column, which is the sum of the "one's" in each row. When completed, the table will look like Table 3-1.

Table 3-1 Majority Table

A	B	C	D	E	#	A	B	C	D	E	#
0	0	0	0	0	0	1	0	0	0	0	1
0	0	0	0	1	1	1	0	0	0	1	2
0	0	0	1	0	1	1	0	0	1	0	2
0	0	0	1	1	2	1	0	0	1	1	3
0	0	1	0	0	1	1	0	1	0	0	2
0	0	1	0	1	2	1	0	1	0	1	3
0	0	1	1	0	2	1	0	1	1	0	3
0	0	1	1	1	3	1	0	1	1	1	4
0	1	0	0	0	1	1	1	0	0	0	2
0	1	0	0	1	2	1	1	0	0	1	3
0	1	0	1	0	2	1	1	0	1	0	3
0	1	0	1	1	3	1	1	0	1	1	4
0	1	1	0	0	2	1	1	1	0	0	3
0	1	1	0	1	3	1	1	1	0	1	4
0	1	1	1	0	3	1	1	1	1	0	4
0	1	1	1	1	4	1	1	1	1	1	5

Then, referring to the table, find every row that has a sum of three. For each of these rows, write the Boolean expression for the combination of the three tanks for that row (the columns that contain a 1). The ten combinations of three empty tanks are: CDE, BDE, BCE, BCD, ADE, ACE, ACD, ABE, ABD, and ABC.

When we write the program for this problem, we can economize on relay contacts in our program. We should keep in mind that simplifying a complex relay structure will save on PLC memory space used by the program. However, if the simplification makes the program difficult for another programmer to read and understand, it should not be simplified. We will simplify by factoring, and then check to see if the ladder diagram is easily readable. We will factor as follows:

$$A(B(C + D + E) + C(D + E) + DE) + B(C(D + E) + DE) + CDE$$

The reader is invited to algebraically expand this expression to verify that all ten of the combinations are covered. Next, we will arbitrarily assign switches A-E to inputs IN1 through IN5 respectively and draw the ladder rung as shown in Figure 3–12.

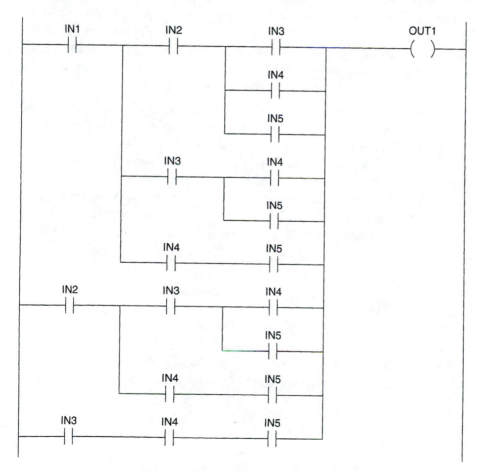

FIGURE 3–12

5-Input Majority Circuit

3-6 Oscillator

With the previous examples, little or no discussion has been made of scan operations or timing. We have merely assumed that when all the conditions were met for a coil to energize, it would do so. However, we must always be aware of the procedure the controller uses to solve the ladder logic diagram. As an example of how an understanding of scanning can benefit us as programmers, let us develop an oscillator. In a PLC program, an oscillator is a coil that turns ON and OFF alternately on each scan. An oscillator can be useful to control things such as math functions and other data manipulation functions that are controlled by a **transitional contact.** A transitional contact is a contact that switches from closed to open or from open to closed. For example, math functions in some controllers only perform their assigned process on the one scan when the control logic switches from open to closed. As long as the control logic remains closed or open, the function will not be performed. To enable the function to occur on an ongoing basis, a transitional contact may be placed in the control logic. This will cause the function to be performed on (in the case of using an oscillator) every other scan because the transitional contact from the oscillator will switch from open to closed on every other scan.

We are now going to get away from the previous process of looking at Boolean equations and electrical circuit diagrams to instead discuss the internal operation of the controller while processing ladder logic. While a good understanding of Boolean logic is essential to understand the process of solving a particular rung, Boolean equations and electrical diagrams will not supply all the tools and understanding you will need to program these devices. By far, experienced programmers rely more on a thorough knowledge of how the controller proceeds to the solution of the ladder and their ingenuity than they do on strict Boolean logic.

Consider the ladder diagram of Figure 3–13. This program uses an internal relay. Recall that internal relays are created by the programmer, can be given any name (in this case, CR1), and are not accessible by terminals on the outside of the PLC. The number of internal relays is limited by the design of the particular PLC being used. The programmer may create only one coil for each internal relay, but may create as many N/O and N/C contacts of each relay as needed. It is important to remember that these relays don't actually exist in a physical sense. Each one is simply a digital bit stored in the PLC.

FIGURE 3–13
Oscillator

The ladder of Figure 3–13 is very simple and consists of only a normally closed (notice normally closed) contact and a coil. The contact and coil have the same number; therefore, if the coil is energized the contact will be open, and if the coil is de-energized the contact will be closed. That is how this configuration functions to provide a transitional contact.

The first thing the controller does when set into operation is perform an I/O update. In the case of this ladder diagram, the I/O update does nothing for us because neither the contact nor the coil are accessible from the outside world (neither is an input nor output). After the I/O update, the controller moves to the **contact logic** portion of the first rung. In this case, this is normally closed contact CR1. The contact logic portion is solved to determine if the coil associated with the rung is to be de-energized or energized. In this case, since the controller has just begun operation, all coils are in the de-energized state. This causes normally closed contact CR1 to be closed (CR1 is de-energized). Since contact CR1 is closed, coil CR1 will be energized. Since this is the only rung of logic in the ladder, the controller will then move on to perform another I/O update. After the update, it will then move on to the first rung (in our example, the only rung) of logic and solve the contact logic. Again, this is one normally closed CR1 contact. However, now CR1 is energized from the last scan, so, contact CR1 will be open. This will cause the controller to de-energize coil CR1. With the ladder diagram solution complete, the controller will then perform another I/O update. Once again, it will return to solve the first rung of logic. The solution of the contact logic will indicate that the contact CR1, on this scan, will be closed because coil CR1 was de-energized when the rung was solved on the last scan. Since the contact is closed, coil CR1 will again be energized. This alternating . . . ON-OFF-ON-OFF-ON . . . sequence will continue as long as the controller is operating. The coil will be ON for one entire scan and OFF for the next entire scan. No matter how many rungs of logic are in the program, for each scan, coil CR1 would be alternating ON for one scan, OFF for one scan, ON for one scan, and so on. For a function that only occurs on an OFF-to-ON contact transition, the transition will occur on every other scan. This is one method of forcing a transitional contact.

In controllers that require such a contact, there is generally a special coil that can be programmed that appears to the controller to switch from OFF to ON on every scan. The rung containing this coil is placed just before the rung requiring the contact. The controller forces the coil containing the transitional contact to an OFF state at the end of the ladder diagram solution so when the rung to be solved is reached, the coil is in the OFF state and must be turned ON. This provides an OFF-to-ON transition on every scan that may be used by the function requiring such a contact.

The understanding of the operation of this oscillator rung will provide an insight into the operation of the controller that will better help the designer to develop programs that use the controller to its fullest extent. The fact that the controller solves each rung one at a time, left-side logic first and right-side coils last, is the reason that a ladder like Figure 3–13 will function as it does. The same circuit, hardwired using an actual relay, would merely buzz after the application of power and perform no useful purpose (unless a buzzer is the desired outcome). It could not be used to energize any other coil since there is no timing to that type of operation. The reason it would only buzz is that in a hardwired system all rungs are solved and the coils set or reset at the same time. It is the timing and sequential operation of the programmable controller that makes it capable of performing otherwise extremely complicated operations.

Let us now add an additional contact to the rung shown in Figure 3–13 to create the gated oscillator in Figure 3–14. The additional contact in this rung is a normally open IN1 contact. If IN1 is OFF at I/O update, normally open IN1 will be open for that scan. If IN1 is ON at I/O update, normally open contact IN1 will be closed for that scan. This provides us with a method of controlling coil CR1's operation. If we want CR1 to provide a transitional (oscillating) contact, we would turn ON IN1. If we want CR1 to be inactive for any reason, IN1 would be turned OFF.

FIGURE 3–14
Gated Oscillator

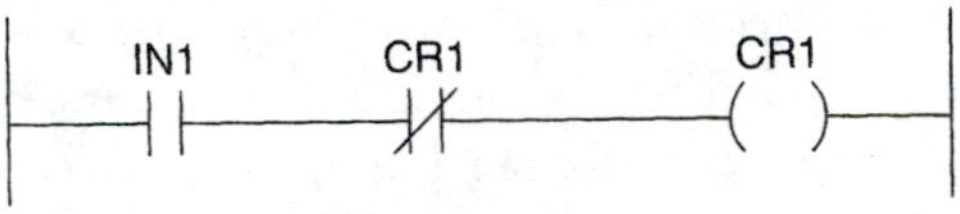

3-7 Holding (also called Sealed, or Latched) Contacts

There are instances when a coil must remain energized after contact logic has been found to be true even if, on successive scans, the logic solution becomes false. A typical application of this would be an ON/OFF control using two separate switches, one to turn the equipment on and one to turn the equipment off. In this case, the coil being controlled by the switches must energize when the ON switch is pressed and remain energized until the OFF switch is pressed. This function is accomplished by developing a rung which contains a **holding contact** (also called a sealing contact or seal-in contact) that will maintain the coil in the energized state until released. Such a configuration is shown in Figure 3–15.

FIGURE 3–15
Holding (or Sealing or Seal-In) Contact

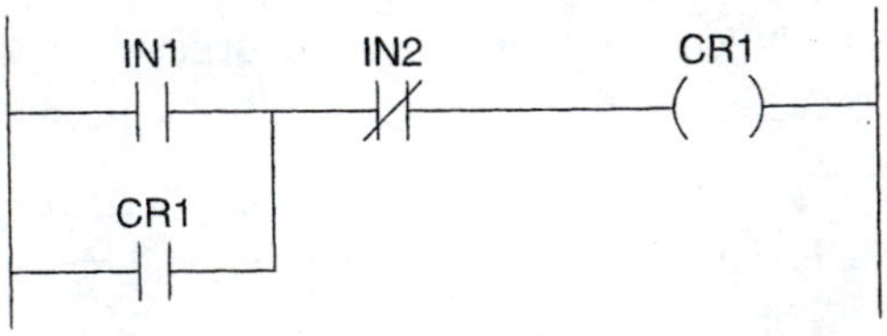

Notice that the contact logic of Figure 3–15 contains three contacts, two normally open and one normally closed. Normally open contact IN1 is defined as the ON switch for our circuit and normally closed contact IN2 is defined as the OFF switch. Notice that when IN1 is turned ON (the normally open contact closes) and IN2 is turned OFF (the normally closed contact is closed), coil CR1 will energize. After CR1 energizes, if IN1 is turned OFF (the normally open contact is open), coil CR1 will remain energized on subsequent scans because normally open contact CR1 will be closed (since coil CR1 would have been energized on the previous scan). Coil CR1 will remain energized until normally closed contact IN2 is opened by turning IN2 ON because when the normally closed contact IN2 opens the contact logic solution will be false. If IN2 is then turned OFF (the normally closed contact closes), coil CR1 will remain de-energized because the solution of the contact logic will be false since normally open contacts IN1 and CR1 will both be open. Normally open contact CR1 is referred to as a holding contact. Therefore, the operation of this rung of logic would be as follows: When the ON (IN1) switch is momentarily pressed, coil CR1 will energize and remain energized until the OFF switch (IN2) is momentarily pressed.

A holding contact allows the programmer to provide for a coil that will hold itself ON after being energized for at least one scan. Such a configuration may be required for an ON/OFF control and on occasions when a fault may occur for only one scan and must be detected at a later time (the coil could be latched ON when the fault occurs). Another name for such a rung is a latch, and the coil is said to be latched ON by the contact associated with the coil.

3–8 Always-ON and Always-OFF Contacts

As programs are developed, there are times when a contact is required that is always ON. In newer PLCs there is generally a coil set aside that meets this requirement. There are, however, some instances where the programmer will have to generate this type of contact in the ladder. For instance, such a contact would be required for a level-triggered (not transition-triggered) arithmetic operation that is to be performed on every scan. Most PLCs require that at least one contact be present in every rung. To satisfy this requirement and have an always true logic, a contact must be placed in the rung that is always true (always closed). There are two ways to produce such a contact. One is to create a coil that is always de-energized and use a normally closed contact associated with the coil. The other is to create a coil that is always energized and use a normally open contact associated with the coil. Figure 3–16 illustrates a ladder rung that develops a coil that is always de-energized.

FIGURE 3–16

Always De-Energized Coil

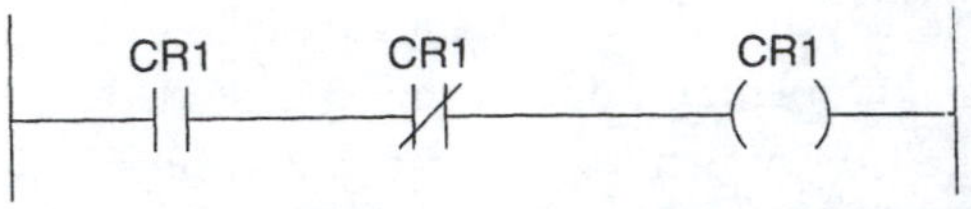

Placing this rung at the top of the program will allow the programmer to use a normally closed contact throughout the ladder anytime a contact is required that is always ON. Notice that coil CR1 will always be de-energized because the logic of normally open CR1 contact AND normally closed CR1 contact can never be true. Anytime a contact is required that is always closed, a normally closed CR1 contact may be used since coil CR1 will never energize.

Figure 3–17 illustrates a rung that creates a coil CR1 that is always energized. Notice that the logic solution for this rung is always true since either normally closed CR1 contact OR normally open CR1 contact will always be true. This will cause coil CR1 to energize at the conclusion of the solution of this rung. This rung must be placed at the very beginning of the ladder to provide for an energized coil on the first scan. Anytime a contact is required that is always closed, a normally open CR1 contact may be used since coil CR1 will always be energized.

FIGURE 3–17

Always Energized Coil

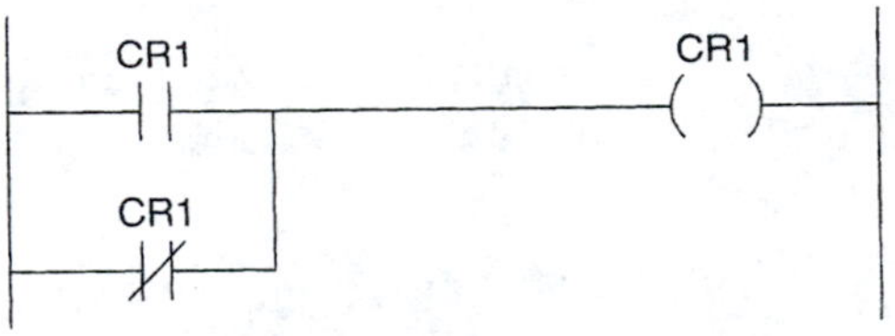

There are advantages and disadvantages to using the always de-energized or always energized coil, and the use depends on the requirement of the software. If an always de-energized coil is used, it will never be energized no matter where in the program the rung is located. An always energized coil will be de-energized until the first time the rung is solved no matter where in the program the rung is located. If there is a program requirement that the coil needs to be de-energized until after the first scan, an always energized coil may be used with the rung placed at the end of the program. This way, the coil would be de-energized until the end of the program where the rung is solved for the first time. After that time, the coil will energize and remain energized until the PLC is turned off.

3-9 Ladder Diagrams Having More Than One Rung

Thus far, we have dealt with either ladder diagrams having only one rung, or multiple rung diagrams that may be rearranged in any desired order without affecting the execution of the program. These have served to solve our problems but leave us limited in the things that can be accomplished. Before moving ahead, let us review the steps taken by the controller to solve a ladder diagram. After the I/O update, the controller looks at the contact portion of the first rung. This logic problem is solved based on the state of the elements as stored in memory. This includes all inputs according to the input image table (the state of all inputs at the time of the last I/O update) and the last known state of all coils (both output and internal). After the contact logic has been solved, the coil indicated on the right side of the rung is either energized (if the solution is true) or de-energized (if the solution is false). Once this is accomplished, the controller moves on to the next rung of logic and repeats the procedure. Once it is set or reset, the coil of a rung will remain in that state until the rung containing the coil is again solved on the next scan. If a contact associated with the coil is used in later rungs, the condition of the contact (energized or de- energized) will be based on the state of the coil at the time of the contact usage. For instance, suppose coil CR1 is energized in rung 3 of a ladder diagram. Later in the ladder, say at rung 25, a normally open contact CR1 is used in the control of a coil. The state of contact CR1 will be energized because coil CR1 was energized in the earlier rung. The normally open energized contact will be closed for the purpose of solving the logic of ladder rung 25. After a rung of logic has been solved and the coil setup accordingly, the controller never "looks" at that rung again until the next scan.

As an aid to further explain these steps, refer to the ladder diagram of Figure 3–18. Notice that the ladder has only one input, IN1, and one output, OUT1. Coil CR1 is referred to as an internal coil. Because it is not accessible from outside the controller as is OUT1, the state of CR1 cannot be readily monitored without being able to see "inside" the machine. In the first rung, IN1 is controlling CR1. In the second rung, a normally open contact of CR1 is used to control the state of OUT1. The solution of the ladder diagram is performed in the following paragraph.

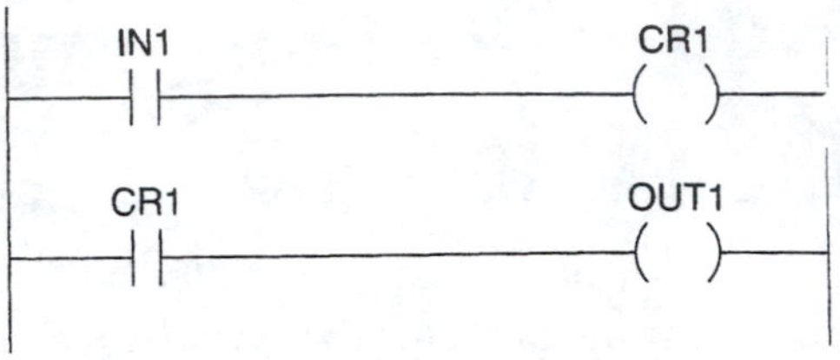

FIGURE 3–18

Two-Rung Ladder Diagram

After I/O update, the controller looks at the contact configuration of rung 1. In this ladder diagram, this is contact IN1. If contact IN1 is closed (energized, since it is normally open), the solution of the contact logic will be true. If contact IN1 is open (de-energized), the solution will be false. Once the contact logic is solved, the controller moves to the coil of rung 1 (CR1). If the contact logic solution is true, coil CR1 will be energized; if the contact logic solution is false, coil CR1 will be de-energized. Notice the phrase "contact logic *is* true or false" in the previous sentence. This is important because after the controller solves the contact logic of a rung, it will not consider it again until the next time the rung is solved (in the next scan). So, if IN1 had been turned ON at the last I/O update, CR1 would be energized after the solution of the first rung. Conversely, if IN1 was OFF at the time of the last I/O update, CR1 would be de-energized after the solution of the first rung. The coil will remain in this state until the next time the controller solves a rung with this coil as the right side, generally on the next scan. After the solution of the first rung is complete, the controller moves to the contact solution of the second rung.

The contact logic of rung 2 consists of one normally open contact, CR1. The condition of this contact at this time depends upon the state of coil CR1 at this time. If coil CR1 is presently energized, contact CR1 will be closed. If coil CR1 is presently de-energized, contact CR1 will be open. After arriving at a true or false solution for the contact logic, the controller will energize or de-energize OUT1 depending on the result. This completes the solution of the entire ladder diagram. The controller then moves on to I/O update. At this update, the output terminal OUT1 will be turned ON or OFF depending upon the state OUT1 was set to in rung 2, and IN1 will be updated to the present state of IN1 at I/O update time. The controller will then move back to the contact logic of rung 1 and start the ladder solution process all over again. The solution of the ladder diagram is a sequential operation. For the purpose of analyzing the operation of the ladder, the states of inputs, internal coils, and output coils can be followed by making a table of states. Such a table follows. The table takes into account all possible combinations of inputs. The table is filled in as each rung is solved. Each line of the table represents the solution of one rung of logic.

	IN1	CR1	OUT1
1	0	0	0
I/O update			
2	1	1	0
3	1	1	1
I/O update			
4	0	0	1
5	0	0	0

Line 1 indicates the condition of inputs and coils at the time the controller is started. IN1 and all coils begin in the OFF condition. Line 2 shows the condition of the elements after the solution of rung 1 assuming that IN1 was ON at the last I/O update. Line 3 shows the condition of the elements after the solution of rung 2. IN1 remains ON for rungs 2 and 3 because if it was ON at I/O update, it will be used as ON for the entire ladder. (The state of all inputs can only change at I/O update.) Lines 4 and 5 show the element states after solving rungs 1 and 2, respectively, assuming that IN1 was turned OFF. Once the timing and sequencing of ladder diagram processing by the controller is better understood, this table approach can be used with each line indicating the results of processing one scan instead of one rung. This can be a much more useful approach especially since the table could become unmanageable in a ladder that has a large number of rungs.

Summary

As we have seen, there are some obvious advantages in using a PLC to perform even the simplest logical control operations. Since the PLC software operating system allows us to internally create multiple N/O and N/C contacts from a single input signal, we are allowed to expand the capabilities of a single input to perform multiple control functions without any increase in the complexity of the wiring surrounding the PLC. This reduction in the wiring complexity is one of the strengths in using a PLC to perform logical control operations. As we will see in subsequent chapters, the capability to internally expand input signals into multiple control operations will be used to an even greater extent and will make the PLC an even more attractive choice for machine control operations.

Review Questions

1. When a relay coil is energized, what is the condition of its normally closed contacts?

2. What is the limitation on the number of contacts associated with a particular relay coil in a PLC program?

3. How is the state of a relay coil represented inside the PLC?

4. If a particular coil is to be an output of the PLC, when is the state of the coil transferred to the output terminals?

5. Draw the ladder logic rung for a normally open IN1 AND'ed with a normally closed IN2 driving a coil CR1.

6. Repeat 5 above but OR IN1 and IN2.

7. What physical changes would be required to system wiring if the PLC system of problem 5 had to be modified to operate as problem 6?

8. Draw the ladder logic rung for a circuit in which IN1, IN2, and IN3 all have to be ON or IN1, IN2, and IN3 all have to be OFF in order for OUT1 to energize.

9. It is desired to implement a switch system similar to a three-way switch system in house wiring; that is, a light may be turned ON or OFF from either of two switches at doors on opposite ends of the room. If the light is turned ON at one switch, it may be turned OFF at the other switch and vice versa. Draw the ladder logic rung that will provide this. Define the two switches as IN10 and IN11 and the output that will control the light as OUT18.

10. Draw the ladder logic rung for an oscillator that will operate only when IN3 and IN5 are both ON or both OFF.

11. The truth table, Table 3–2, shows all the possible combinations for the inputs IN1, IN2, and IN3 shown in Figure 3–19 with each row being a unique combination of the inputs being ON or OFF (a 1 indicates the input is in an ON condition and a 0 indicates an OFF condition). Complete the table by writing the state (1 or 0) of OUT1 for each input combination. Reminder: A N/C contact is closed when its input is OFF.

Table 3–2 Problem 11 Truth Table

IN1	IN2	IN3	OUT1
0	0	0	
0	0	1	
0	1	0	
0	1	1	
1	0	0	
1	0	1	
1	1	0	
1	1	1	

FIGURE 3–19
Ladder Diagram for Problem 11

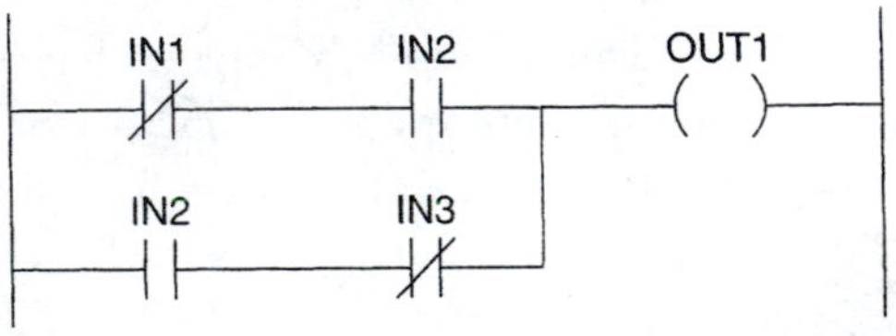

12. Draw a ladder diagram that will cause OUT2 to be ON under the conditions shown in the truth table, Table 3-3. (Hint: It may help to write a Boolean expression.) Be sure to simplify the ladder so that it uses the minimal number of relay contacts.

Table 3-3 Problem 12 Truth Table

IN1	IN2	IN3	OUT2
0	0	0	0
0	0	1	0
0	1	0	1
0	1	1	1
1	0	0	1
1	0	1	0
1	1	0	0
1	1	1	0

Advanced Programming Techniques

Objectives

Upon completion of this chapter, you will know:

- how to take advantage of the order of program execution in a PLC to perform unique functions.

- how to construct fundamental asynchronous and clocked flip flops in ladder logic including the R-S, T, D, and J-K types.

- how the one shot operates in a PLC program and how it can be used to edge-trigger flip flops.

- how to use uni-directional and bi-directional counters in PLC programs.

- how sequencers function and how they can be used in a PLC program.

- how timers operate and the difference between retentive and non-retentive timers.

Introduction

In addition to the standard logical operations that a PLC can perform, seasoned PLC programmers are aware that, by taking advantage of some of the unique features and characteristics of a PLC, some very powerful operations can be performed. Some of these are operations that would be very difficult to create in hardwired relay logic but are relatively simple in PLC ladder programs. Many of the program segments in this chapter are rather "cookbook" by nature. The reader should not concentrate on memorizing these programs, but instead, learn how they work and how they can be best applied to solve programming problems.

4-1 Ladder Program Execution Sequence

Many persons new to PLC ladder logic programming may tend to think that, because a PLC executes its program synchronously (i.e., from left to right, top to bottom) instead of asynchronously (i.e., each relay operates whenever it receives a signal), it is a hindrance to the programming task. However, after gaining some experience with programming PLCs, the programmer learns how to use this to their advantage. In this chapter, we will see several useful program segments that do this. Keep in mind that the order of the rungs in these programs is critical. If the rungs are rearranged in another order, it is likely that these programs will not operate properly.

4-2 Flip Flops

As we will see in the following several sections, a few of the circuits you learned to use in digital electronics can be developed for use in a ladder diagram. The ones we will study are the R-S, D, T, and J-K Flip Flops and the One Shot. The One Shot will supply the clock (trigger) pulse for the D, T, and J-K Flip Flops. These are functions that can be very useful in a control system and tend to be more familiar to the reader, especially one who has previously studied digital circuits.

4-3 R-S Flip Flop

The R-S Flip Flop is the most basic of the various types. It has two inputs, an R (reset) and an S (set). Turning ON the R input resets the flip flop and turning ON the S input sets the flip flop. As you may recall from the study of this type of flip flop, the condition where R and S are both ON is an undefined state and should be avoided. The truth table for an R-S Flip Flop is shown in Table 4–1.

Table 4-1 Truth Table for R-S Flip Flop

R	S	Q	Q'
0	0	Q	Q'
0	1	1	0
1	0	0	1
1	1	X	X

For the purpose of our discussion, a 1 in the table indicates an energized condition for a coil or contact. An X in the table indicates a "don't care" condition. The ladder diagram for such a circuit is shown in Figure 4–1.

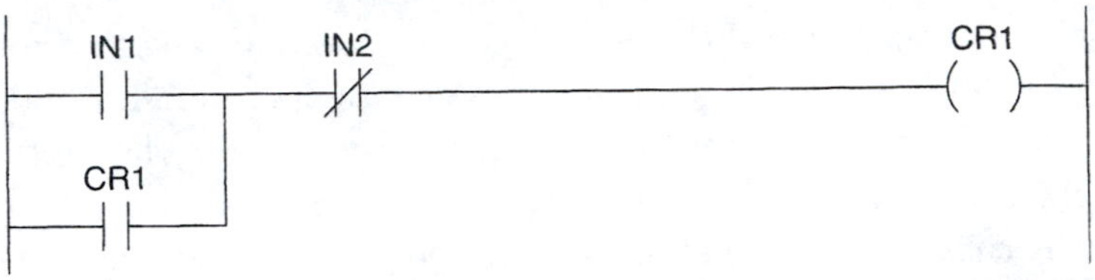

Notice in Figure 4–1 that if IN2 energizes, coil CR1 will de-energize, and if IN2 de-energizes and IN1 energizes, coil CR1 will energize. Once CR1 energizes, contact CR1 will close, holding coil CR1 in an energized state (even after IN1 de-energizes) until IN2 energizes to reset the coil. Relating to the truth table of Table 4–1, notice that IN1 is the S input and IN2 is the R input to the flip flop. Also notice in the truth table that both a Q and Q′ outputs are indicated. If an inverted Q signal is required from this ladder to be used in another rung, one only needs to make the contact a normally closed CR1 (a normally closed contact is open when its coil is energized).

4-4 One Shot

As with the oscillator covered in a previous chapter, the **one shot** has its own definition in ladder logic. In digital electronics, a one shot is a monostable multivibrator with an output that is ON for a predetermined length of time. This time is adjusted by selecting the proper timing components. A one shot in ladder logic is a coil that is ON for only one scan each time it is triggered. The length of time the one shot coil is ON depends on the scan time of the PLC. The one shot can be triggered by an outside input to the controller or from a contact associated with another coil in the ladder diagram. It can also be a coil that energizes for one scan automatically at startup. It is this type we will study first, followed by the externally triggered type.

A one shot that comes ON for one scan at program startup is shown in Figure 4–2.

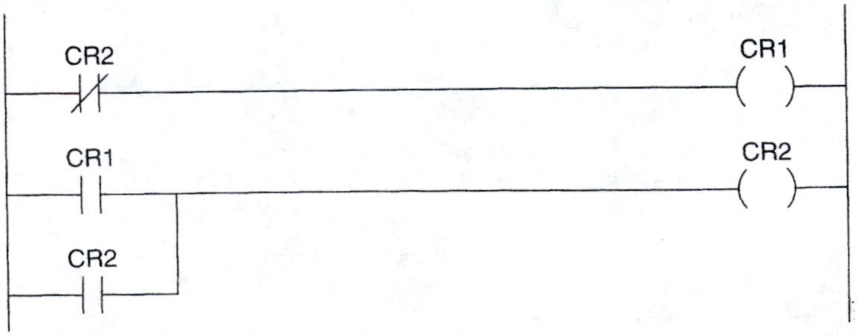

FIGURE 4–2

Automatic One
Shot CR1 = Q

This one shot consists of two rungs of logic that are placed at the beginning of the ladder diagram. Coil CR1 is the one shot coil. By analyzing the first rung, it can be seen that, since all coils are OFF at the beginning of operation, normally closed contact CR2 will be closed (because coil CR2 is de-energized). This will result in coil CR1 being energized in the first rung of the ladder. The program execution then moves to the second rung, which contains the remaining part of the one shot. When the controller solves the contact logic of this rung, normally open contact CR1 will be closed (CR1 was energized in the first rung) and normally open contact CR2 will be open (CR2 is still de-energized from startup). Since contact CR1 is closed, coil CR2 will be energized in the second rung. The PLC will then proceed with solutions of the remaining rungs of the ladder diagram. After I/O update, the controller will return to rung 1. At this time contact CR2 will be open (coil CR2 was energized on the previous scan), the solution to the contact logic will be false, and coil CR1 will be de-energized. When the PLC solves the logic for rung 2, the contact logic will be true because even though coil CR1 is de-energized making contact CR1 open, coil CR2 is still energized from the previous scan. This makes contact CR2 closed, which will keep coil CR2 energized for as long as the controller is operating. Coil CR2 will remain closed because each time the controller arrives at the second rung, contact CR2 will be closed due to coil CR2 being energized from the previous scan. The result of this ladder is that coil CR1 will energize on the first scan of operation, de-energize on the second scan, and remain de-energized for the remaining time the controller is in operation. However, coil CR2 will be de-energized until the first solution of rung 2 on the first scan of operation, then energize and remain energized for the remaining time the controller is in operation. Each time the controller is turned OFF, then turned back ON, this sequence will take place. This is because the controller resets all coils to their de-energized state at the onset of operation. If any operation in the ladder requires that a contact be closed or open for the first scan after startup, this type of network can be used. Coil CR1 will remain closed for the first scan of operation regardless of the time required.

Notice, however, that we have two coils, CR1 and CR2, that are complements of each other. Whenever one is ON, the other is OFF. In ladder logic, this is redundant because if we need the complement of a coil, we merely need to use a normally closed contact from the coil instead of a normally open contact. For this reason, a simpler type of one shot that is in one state for the first scan and the other state for all succeeding scans can be used. This is a coil that is OFF for the first scan and ON for all other scans. Any time a contact from this coil is required, a normally closed contact may be used. This contact would be closed for the first scan (the scan that the coil is de-energized) and open for all other scans (the coil is energized). Such a ladder diagram is shown in Figure 4–3.

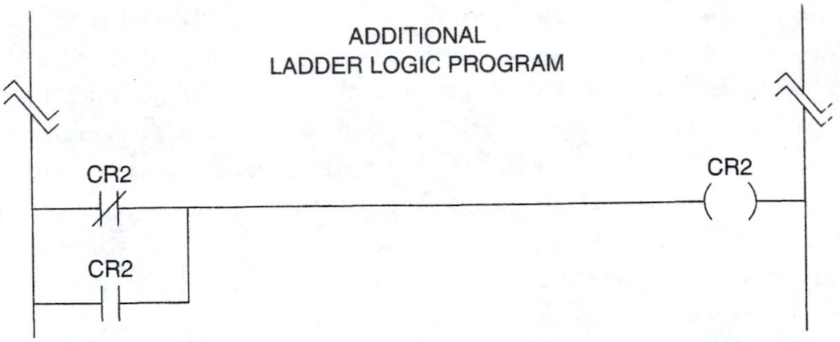

FIGURE 4–3

*One Shot Ladder
Diagram
CR2' = Q*

FIGURE 4–3 caption content shown in figure:

```
                        ADDITIONAL
                   LADDER LOGIC PROGRAM


        CR2                                              CR2
       —|/|—————————————————————————————————————————————( )—
        CR2
       —| |—
```

This one shot only consists of one rung of logic, which is *placed at the end of the ladder diagram.* Notice that the contact logic for the rung includes two contacts, one normally open and one normally closed. Since both contacts are for the same coil (CR2), no matter what the status of CR2, the contact logic will be true. This can be proven by writing the Boolean expression for the contact logic and reducing. An expression that contains a signal or its complement is always true $(A + \overline{A} = 1)$. This means that CR2 will be OFF until the controller arrives at the last rung of logic. When the last rung is solved, the result will be that CR2 is turned ON. Each time the controller arrives at the last rung of logic on each scan, the result will be the same. CR2 will be OFF for the first scan and ON for every scan thereafter until the controller is turned OFF and ON again. Any rung that requires a contact that is ON for the first scan only would include a normally closed CR2 contact. Any rung that requires a contact that is OFF for the first scan only would contain a normally open CR2 contact.

Many of the newer PLCs provide a first scan one shot as a pre-programmed function. In this case, it is not necessary to construct the one shot since it is provided by the PLC firmware. This special function internal relay is usually named **first scan, power-up scan,** or some other similar name.

The other type of one shot mentioned is triggered from an external source such as a contact input from inside or outside the controller. We will study one triggered from an input contact, although it could also be triggered from an internal relay contact. The ladder diagram for such a one shot is shown in Figure 4–4.

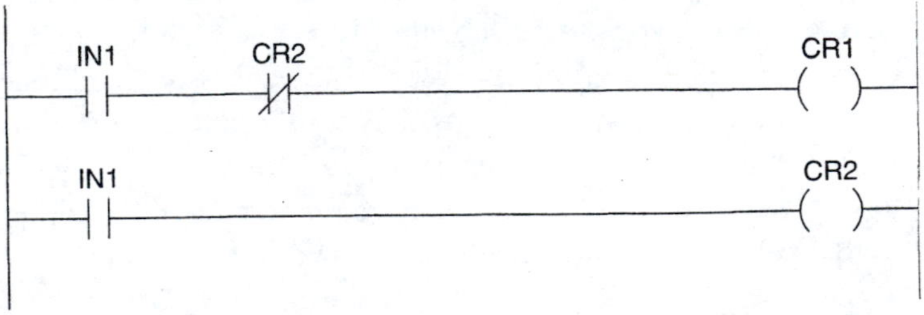

FIGURE 4–4

*Externally
Triggered One Shot
CR1 = Q*

```
        IN1        CR2                                    CR1
       —| |————————|/|——————————————————————————————————( )—
        IN1                                               CR2
       —| |—————————————————————————————————————————————( )—
```

Notice that this type of one shot is also composed of two rungs of logic. In this case, the portion of the ladder that needs the one shot contact is placed after the two rungs. In operation, with IN1 de-energized (open), the two coils, CR1 and CR2, are both de-energized. On the first scan after IN1 is turned ON, CR1 will be energized in the first rung and CR2 in the second. On the second scan after IN1 is turned ON, CR1 will de-energize due to the normally closed CR2 contact in the first rung. CR2 will remain ON for every scan that IN1 is ON. The result is that coil CR1 will energize for only the first scan after IN1 turns ON. After that first scan, coil CR1 will de-energize. The system will remain in that state, CR1 OFF and CR2 ON, until the scan after IN1 turns OFF. On that scan, coil CR1 will remain de-energized in the first rung, and coil CR2 will remain de-energized in the second rung. This makes the system ready to again produce a one shot signal on the next IN1 closure, with coil CR1 controlling the contact that will perform the function. If a contact closure is required for the function requiring the one shot signal, a normally open CR1 contact may be used. If a contact opening is required, a normally closed CR1 contact may be used. This type of one shot function is helpful in implementing the next types of flip flops, the D, T, and J-K Flip Flops. The D Flip Flop may be designed with or without the one shot while the T and J-K Flip Flops require a clock (trigger) signal. In this case, we will use an input clock and a one shot to produce the single scan contact closure each time the input is turned ON.

4–5 D Flip Flop

A D Flip Flop has two inputs—the D input and a trigger or clock input. In operation, the state of the D input is transferred to the Q output of the flip flop at the time of the clock pulse. The truth table for a D Flip Flop is shown in Table 4–2.

Table 4–2 Truth Table for D Flip Flop

D	CL	Q_n	Q_{n+1}
0	0	X	Q_n
0	1	X	0
1	0	Q	Q_n
1	1	X	1

In Table 4–2, the column labeled Q_n contains the state of the flip flop Q output prior to the clock, and the column labeled Q_{n+1} contains the state of the Q output of the flip flop after the application of the clock. An X indicates a "don't care" situation. A 1 in the CL column indicates that the clock makes a 0 to 1 to 0 transition.

A ladder D Flip Flop is shown in Figure 4–5. The D Flip Flop shown is a one rung function with two inputs, IN1 and IN2, and one coil, CR1. In this case, IN1 is the D input and IN2 is the clock. The clock for the D Flip Flop can be either a normal contact closure (a level trigger) or a single scan (or edge trigger) contact closure from a one shot. We will begin with a discussion of the level triggered D Flip Flop. Let us assume that the controller has just been set into operation and that all coils are de-energized. Also, at this time we will let IN1 and IN2 be OFF, resulting in normally open contacts IN1 and IN2 being open and normally closed IN2 contact being closed. In this state, coil CR1 will remain de-energized on every scan. Now, let IN1 (the D input) turn ON. Each time the controller solves the rung in this state, the upper branch of logic (IN1 and IN2 contacts) will be false because IN2 remains OFF, and the lower branch of contacts (CR1 and IN2) will be false because CR1 is de-energized. This means that coil CR1 will remain de-energized. If IN2 is now turned ON, the upper branch of contacts will become true because IN1 and IN2 will both be ON. The result is that coil CR1 will energize on the first scan after IN2 turns ON and will remain energized as long as IN2 is ON since IN2 is the clock signal for the flip flop. When the clock signal IN2 is turned OFF, coil CR1 will remain in the energized state. This is because the lower branch (N/O CR1 and N/C IN2) will be true (CR1 is energized and IN2 is OFF). If IN2 (the clock) turns ON and OFF again and IN1 is still ON, the same events will occur; the upper branch of contacts will be true, which will hold coil CR1 energized while IN2 is ON and the lower branch of contacts will be true and hold CR1 energized when IN2 turns OFF. If the D input to the flip flop is true, we want the Q output to stay in a 1 state after the clock. Now consider the condition when the D input is a 0. On the first scan after IN2 (the clock) turns ON, if IN1 (D) is OFF, coil CR1 will de-energize. This is because both branches of contacts will be false—the upper because IN1 is OFF and the lower because IN2 is ON. On subsequent scans, coil CR1 will remain de-energized because, again, both branches of contacts will be false—the upper with IN2 OFF and the lower with CR1 OFF.

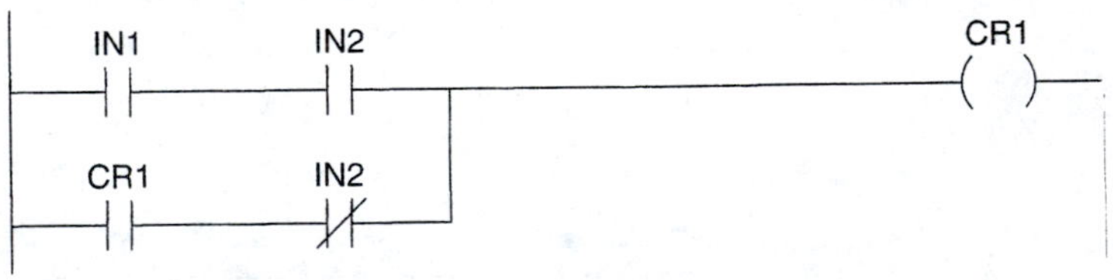

FIGURE 4–5

*Ladder Diagram
for D Flip Flop
IN1 = D,
IN2 = Clock*

Next, consider the operation of a D Flip Flop having D input IN1 with a single scan clock that is initiated from contact IN2. For this case, the truth table in Table 4–2 is still valid. The ladder diagram for such a system is shown in Figure 4–6.

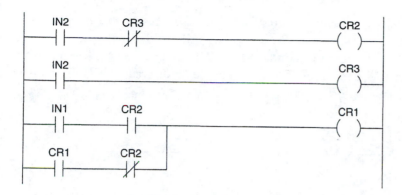

FIGURE 4–6

Ladder Diagram for D Flip Flop with Single Scan Trigger, IN1 = D, IN2 = Clock

Notice that the D Flip Flop has the same appearance as before except for a different contact for the clock. Previously, the clock was an input contact, now it is an internal coil CR2. The first and second rungs are the externally triggered one shot of the type in Figure 4–4. The input contact in this example has been changed to IN2 and the one shot coil is now CR2. Each time IN2 is turned on, coil CR2 will energize for one scan only. The action on the flip flop will be the same as if the clock input (IN2) of Table 4–2 were turned on for one scan only. Keep in mind that the trigger contact that initiates the one shot function could be a contact from a coil within the ladder. This would allow the ladder itself to control the flip flop operation from inside the program.

4-6 T Flip Flop

The T Flip Flop also has two inputs. The clock input performs the same function as the D Flip Flop in that it initiates the flip flop action. Unlike the D Flip Flop, the second input to a T Flip Flop (the T input) enables or disables a toggle operation. A T Flip Flop will remain in its present state upon the application of a clock signal if the T input is a 0. If the T input is a 1, the flip flop will toggle to the opposite state upon application of a clock signal. The truth table for this flip flop is shown in Table 4–3.

Table 4–3 Truth Table for T Flip Flop

T	CL	Q_n	Q_{n+1}
0	0	X	Q_n
0	1	X	Q_n
1	0	Q	Q_n
1	1	Q	Q_n'

A 1 in the CL (clock) column indicates that the clock makes a 0 to 1 to 0 transition. The Q_n column is the state of the flip flop prior to the application of the clock and the Q_{n+1} column is the state of the flip flop after the clock. An X in the table indicates a "don't care" condition. The ladder diagram of a T Flip Flop is shown in Figure 4–7.

FIGURE 4–7

Ladder Diagram
for a T Flip Flop
IN1 = T,
IN2 = Clock

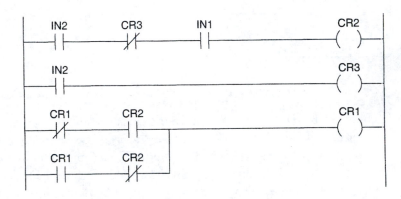

The T Flip Flop is formed by three rungs which include a toggling coil (CR1) and a one shot. The one shot is composed of the first and second rungs. The one shot portion of the T Flip Flop ladder is triggered by IN2, and there is an additional normally open contact (IN1) in the first rung. The purpose of the normally open IN1 contact is to provide the T input to the flip flop network. Remember that the T input controls whether or not the flip flop will toggle. If T is a 1 the flip flop will toggle, and if T is a 0 the flip flop will not toggle. The one shot will not trigger if IN1 is a zero because IN1 in rung 1 will prevent the contact logic from being true. As a result, the coil CR1 in rung 3 will not toggle if IN1 does not enable the triggering of the one shot. The contact logic for rung 3 has two branches, one containing the AND combination of a normally closed CR1 contact and a normally open CR2 contact. The lower branch contains the AND combination of a normally open CR1 contact and a normally closed CR2 contact. The two AND contact combinations are OR'ed together to form the total contact logic for the toggle coil CR1. If IN1 and IN2 are both true at the I/O update, the one shot will trigger just as in our previous discussion of one shot operation. If this is the case, one shot coil CR2 will be energized for only the first scan after that I/O update. If IN1 is not true, one shot coil CR2 will not energize. Let us assume that toggle coil CR1 is de-energized and that one shot coil CR2 has been enabled and is ON for the one scan presently being executed. When program execution arrives at rung 3 and solves the contact logic, the upper branch will be true because coil CR1 is de-energized making normally closed contact CR1 closed, and the normally open CR2 contact will be closed (CR2 is energized for this scan). This will result in the controller energizing coil CR1. On the second and all other scans after IN2 turns ON, coil CR2 will be de-energized. On these scans, when the controller arrives at the third rung, the lower branch of contact logic will be true. This is because CR1 is energized and CR2 is de-energized. IN2 must turn OFF and then ON again for the toggle to operate once more. When this occurs, one shot coil CR2 will again be ON for the first scan after IN2 turns ON (assuming that IN1 was ON at the time). When the controller solves the contact logic of rung 3 on this scan, the upper branch will be false because CR1 is energized and the lower branch will be false because one shot coil CR2 is energized. This will cause toggle coil CR1 to de-energize. On the second and all other scans after IN2 turns ON, both branches will still be false; the upper because coil CR2 is de-energized, and the lower because coil CR1 is de-energized. The rung will continue to be solved with this result until the one shot coil CR2 is again energized for the one scan after IN2 turns ON with IN1 ON.

4-7 J-K Flip Flop

The truth table for the J-K Flip Flop is shown in Table 4–4.

Table 4-4 Truth Table for J-K Flip Flop

J	K	CL	Q_n	Q_{n+1}
0	0	1	Q_n	Q_n
0	1	1	X	0
1	0	1	X	1
1	1	1	Q_n	Q_n'
X	X	0	Q_n	Q_n

For this table, an X in any block indicates a "don't care" condition, a 1 in the CL (clock) column indicates the clock makes a 0 to 1 to 0 transition, and a 0 in the CL column indicates no clock transition. The Q_n column contains the flip flop state prior to the application of a clock, and the Q_{n+1} column contains the flip flop state after the clock. The ladder diagram for a J-K Flip Flop in which IN1 = J, IN2 = K, and IN3 = CL is shown in Figure 4–8.

FIGURE 4–8

Ladder Diagram
for a J-K Flip Flop
IN1 = J,
IN2 = K,
IN3 = Clock

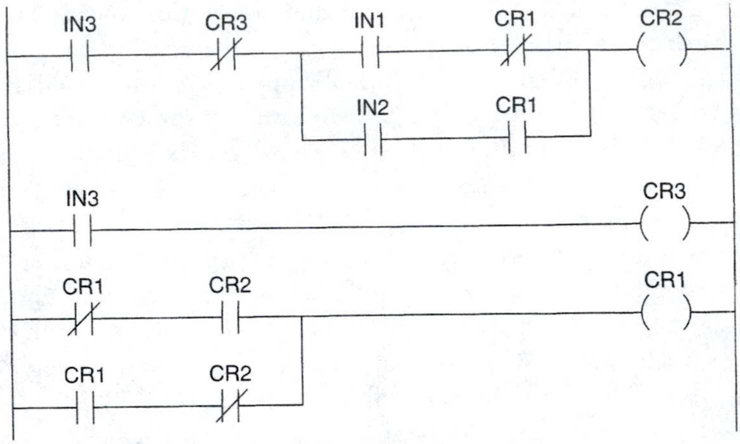

Note that the network in Figure 4–8 is a T Flip Flop (in rung 3) that is triggered by a special one shot located in rungs 1 and 2. It is this special one shot that makes the T Flip Flop perform as a J-K Flip Flop. Referring to Table 4–4, notice that we wish to have the flip flop toggle under either of two input conditions; when J = 1 and Q_n = 0, then Q should toggle to Q_{n+1} = 1, OR when K = 1 and Q_n = 1, then Q should toggle to Q_{n+1} = 0. Therefore, writing this as a Boolean expression, we can say that we wish to have Q toggle under the condition JQ′ + KQ. As the J-K Flip Flop program in Figure 4–8 is constructed, Q will toggle every time the one shot is triggered. Therefore, we wish to trigger the one shot under the condition JQ′ + KQ. If we substitute our component names IN1, IN2, and CR2 for J, K, and Q, respectively, we get the expression IN1 · CR2′ + IN2 · CR2.

If we implement this expression using ladder contact logic, the ladder portion would be as shown in Figure 4–9. Referring back to Figure 4–8, note that this contact configuration is located in the first rung of the ladder and it controls the triggering of the one shot from IN3. The result is that the ladder diagram of Figure 4–8 will function as a J-K Flip Flop with CR1 providing the Q output.

FIGURE 4–9

Contact Logic
Required to
Implement T = KQ + JQ′

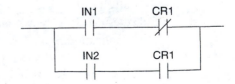

4-8 Counters

A **counter** is a special function included in the PLC program language that allows the PLC to increment or decrement a number each time the control logic for the rung switches from false to true. In its simplest form, this special function generally has two control logic lines—one that causes the counter to count each time the count control becomes true and one that causes the counter to reset when the reset control line is true. A typical counter is shown in Figure 4–10.

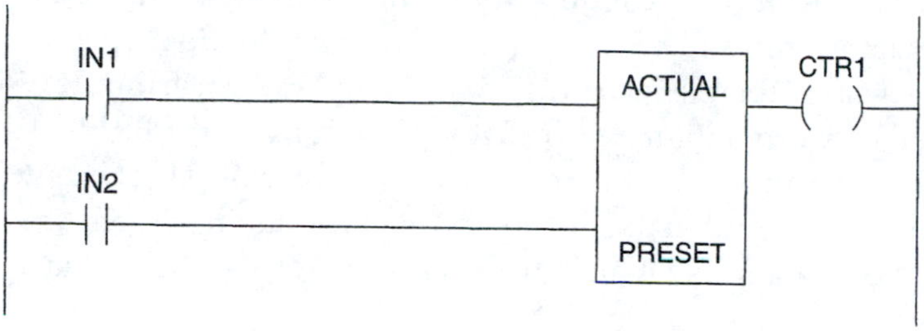

FIGURE 4–10

Counter

Notice that this special function has two control lines—one containing a normally open contact IN1 and one containing normally open contact IN2. The counter itself has a coil associated with it that is numbered CTR1. In some PLC programming languages, the coil is contained within the rectangular box of the timer. Notice too, that inside the function block are two labels, **ACTUAL** and **PRESET**. These ACTUAL and PRESET items contain numbers. The PRESET value is the maximum count allowed for the counter. This number may be held as a constant value in permanent memory or as a variable in a **holding register.** A holding register is a memory location in RAM that may be altered as required. The programmer would use a holding register for the PRESET value of the counter if the maximum count value needed to change depending upon program operation, such as in a program that needed to count items placed in a box. If different size boxes were used, depending upon the product and quantity to be shipped, the counter maximum may need to change. Although the PRESET value may reside in either a holding register or RAM, the ACTUAL value is always maintained in a RAM location because it is the present value of the counter. As the counter counts, this value must change. It is this value compared to the PRESET value that the PLC uses to determine if the counter is at its maximum value. As the ACTUAL value increases, it is compared to the PRESET value. When the ACTUAL value is equal to the PRESET value, the counter will stop counting and the coil associated with the counter (in this case CTR1) will be energized.

In our example in Figure 4–10, contacts IN1 and IN2 control the counter. The top line, containing IN1, is referred to as the **count line.** The lower control line, containing IN2, is referred to as the **reset line.** Note that with some PLC manufacturers the two input lines are the reverse of that shown in Figure 4–10, with the RESET line on top and the COUNT line below. In operation, if IN2 is closed the counter will be held in the reset condition; that is, the ACTUAL value will be set to zero whether IN1 is open or closed. As long as the reset line is true, the ACTUAL value will be held at zero regardless of what happens to the count line. If the RESET LINE is opened, the counter will be allowed to increment the ACTUAL value each time the count control line switches from false to true (OFF to ON). In our example, this will occur each time IN1 switches from open to closed. The counter will continue to increment the ACTUAL value each time IN1 switches from open to closed until the ACTUAL value is equal to the PRESET value. At that time, the counter will stop incrementing and the ACTUAL value and coil CTR1 will be energized. If at any time during the counting process the RESET control line containing IN2 is made to switch to true, the ACTUAL value will be reset to zero.

Different PLC manufacturers handle counters in different ways. Most counters operate as described previously. Another approach taken in some cases is to reset the ACTUAL value to the PRESET value (rather than reset it to zero) and decrement the ACTUAL value toward zero. In this case, the coil associated with the counter is energized when the ACTUAL value is equal to zero rather than when it is equal to the PRESET value.

Some manufacturers have counters that are constructed using two separate rungs. These have an advantage in that the reset rung can be located anywhere in the program and does not need to be located immediately following the count rung. Figure 4–11 shows a counter of this type. In this sample program, note that N/O

FIGURE 4–11

*Two-Rung
Counter and
Output Rung*

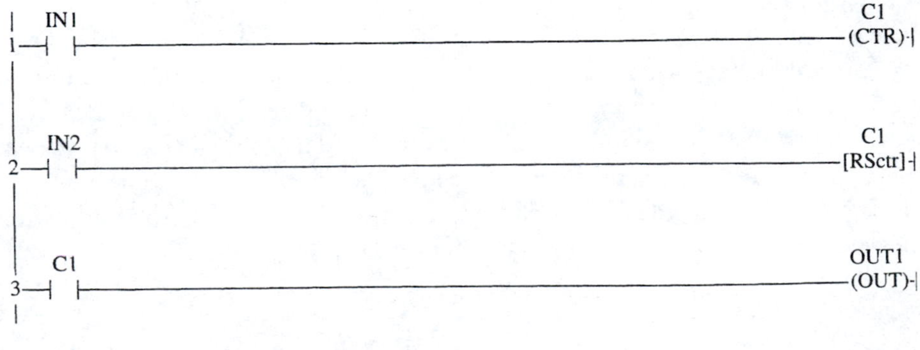

IN1 in rung 1 causes the counter C1 to increment (or decrement, if it is a down counter), and N/O IN2 in rung 2 causes the counter C1 to reset to zero (or reset to the preset value if it is a down counter). Rung 3 has been added to illustrate how a counter of this type can be used. Contact C1 in rung 3 is a contact of counter C1. It is energized when counter C1 reaches its preset value (if it is a down counter, it will energize when C1 reaches a count of zero). The result is that output OUT1 will be energized when input IN1 switches ON a number of times equal to the preset value of counter C1.

In some cases, it is convenient to have a counter that can count in either of the two directions, called a **bi-directional counter.** For example, in a situation where a PLC needs to maintain a running tally of the total number of parts in a que where parts are both entering and exiting the que, a bi-directional counter can be incremented when a part enters and decremented when a part exits the que. Figure 4–12 shows a bi-directional counter, C2, which has three inputs and consists of three rungs. Rung 1 controls the counting of C2 in the up direction, rung 2 controls C2 in the down direction, and rung 3 resets C2.

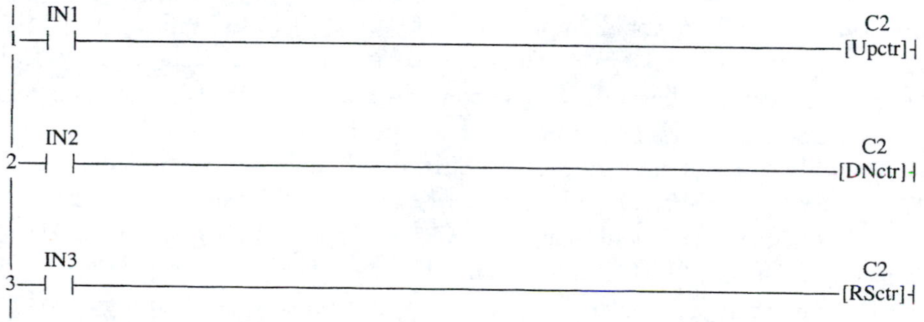

```
|   IN1                                                    C2
1 --| |-------------------------------------------------[Upctr]-|


|   IN2                                                    C2
2 --| |-------------------------------------------------[DNctr]-|


|   IN3                                                    C2
3 --| |-------------------------------------------------[RSctr]-|
```

FIGURE 4–12
Up/Down Counter

4-9 Sequencers

Some machine control applications require that a particular sequence of events occur and, with each step of the controller, a different operation be performed. The programming element for this type of control is called a **sequencer.** For example, the timer in a washing machine is a mechanical sequencer that has the machine perform different operations (fill, wash, drain, spin) in a predetermined sequence. Although a washing machine timer is a timed sequencer, sequencers in a PLC are not necessarily timed. An example of a non-timed sequencer is a garage door opener. It performs the sequence . . . up, stop, down, stop, up, stop. . . with each step in the sequence being activated by a switch input or remote control input.

PLC sequencers are fundamentally counters with some extra features and some minor differences. Counters will generally count to either their preset value (in the case of up counters) or zero (for down counters) and stop when they reach their terminal count. However, sequencers are circular counters; that is, they will "roll over" (much like an automobile odometer) and continue counting. If the sequencer is of the type that counts up from zero to the preset, on the next count pulse after reaching the preset it will reset to zero and begin counting up again. If the sequencer is of the type that counts down, on the next count pulse after it reaches zero it will load the preset value and continue counting down. All sequencers have a reset input that resets them either to zero (for the types that count up) or to the preset value (for the types that count down). As with counters, some PLC manufacturers provide sequencers with a third input (usually called **UP/DN**) that controls the count direction. These are called **bi-directional sequencers** or **reversible sequencers.** Alternately, other bi-directional sequencers have separate count up and count down inputs.

Unlike counters, sequencers have contacts that actuate at any specified count of the sequence. For example, if we have an up-counting sequencer SEQ1 with a preset value of 10, and we would like to have a rung switch ON when the sequencer reaches a count of 8, we would simply put a N/O contact of SEQ1 = 8 (or SEQ1:8) in the rung. For this contact, when the sequencer is at a count of 8, the contact will be ON. The contact will be OFF for all other values of sequencer SEQ1. In our programs, we are allowed as many contacts of a sequencer as desired of either polarity (N/O or N/C) and of any sequence value. If, for example, we would like our sequencer, SEQ1, to switch on an output OUT1 whenever the sequencer is in count 3 or 8 of its sequence, we would simply connect N/O contacts SEQ1:3 and SEQ1:8 in parallel to operate OUT1. This is shown in Figure 4–13. In rung 1, N/O contact IN1 advances the sequencer SEQ1 each time the contacts close. In rung 2, N/O contact IN2 resets SEQ1 when the contact closes. In rung 3, output OUT1 is energized when the sequencer SEQ1 is in either state 3 or state 8.

FIGURE 4–13

Sequencer and Output Rung

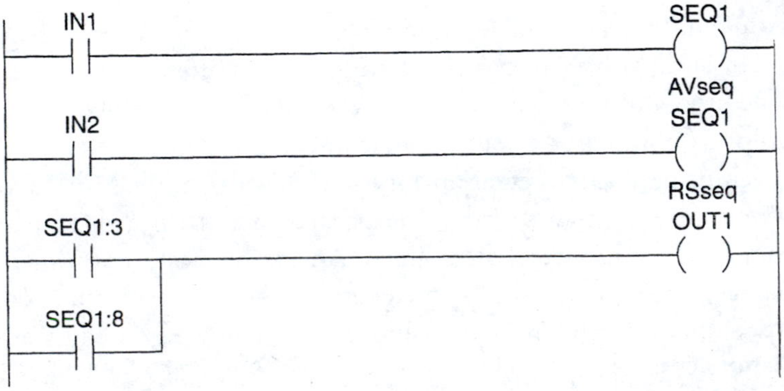

4-10 Timers

A **timer** is a special counter ladder function that allows the PLC to perform timing operations based on a precise internal clock, generally 0.1 or 0.01 seconds per clock pulse. Timers usually fall into two different categories depending on the PLC manufacturer. These are **retentive** and **non-retentive timers.** A non-retentive timer is one that has one control line; that is, the timer is either timing or it is reset. When this type of timer is stopped, it is automatically reset. This will become more clear as discussion of timers continues. The retentive timer has two control lines—count and reset. This type of timer may be started, stopped, and then restarted without resetting. This means that it may be used as a totalizing timer by simply controlling the count line. Independent resetting occurs by activating the reset control line. At the beginning of this section, it was stated that a timer is a special counter. The timing function is performed by allowing the counter to increment or decrement at a rate controlled by the internal system clock. Timers typically increment or decrement at 0.1-second or 0.01-second rates depending upon the PLC manufacturer. In many PLCs, the time increment is user selectable, with the 0.1s incrementing counters called **standard timers** and the 0.01s incrementing timers called **high-speed timers.**

An example of a non-retentive timer is shown in Figure 4–14. Notice that this timer has only one control line containing normally open contact IN1. Also notice that, like the counter, there are two values, ACTUAL and PRESET. These values are, as with the counter, the present and final values for the timer. In Figure 4–14, while the control line containing IN1 is false (IN1 is open), the ACTUAL value of the timer is held reset to zero. When the control line becomes true (IN1 closes), the timer ACTUAL value is incremented each 0.1 or 0.01 second. When the ACTUAL value is equal to the PRESET value, the coil associated with the timer (in this case TIM1) is energized and the ACTUAL value incrementing ceases. The PRESET value must be set so that the timer counter ACTUAL value will increment from zero to the PRESET value in the desired time. For instance, suppose a timer of 5.0 seconds is required using a 0.1 second rate timer. The PRESET value would need to be 50 for this function since it would take 5.0 seconds for the counter to count from zero to 50 utilizing a 0.1 second clock (50 × 0.1 second = 5.0 seconds). If a 0.01 second clock were available, the PRESET value would need to be 500.

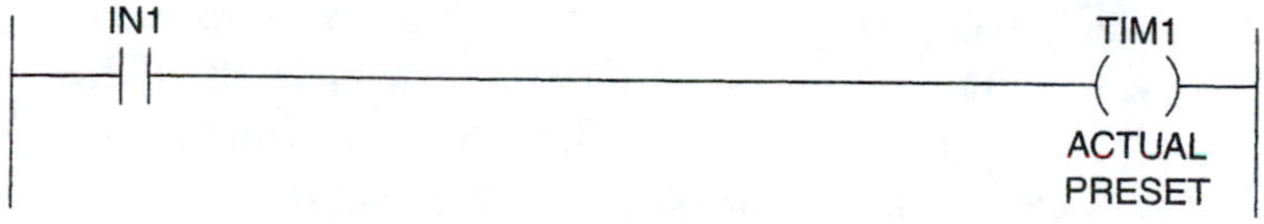

FIGURE 4–14

Non-Retentive Timer

An example of a retentive timer is shown in Figure 4–15. This type of timer looks more like the counter discussed earlier, and the two control lines operate in much the same manner as the counter in that the lower line is the reset line. Note in Figure 4–15 that, in addition to the timer box, there is a TIM1 coil to the right. This is the timer symbol used by some PLC manufacturers. There are several other variations of the timer symbol, some using the box and not the coil and others using the coil and not the box. In any case, all of them perform as a retentive timer.

FIGURE 4–15

Retentive Timer

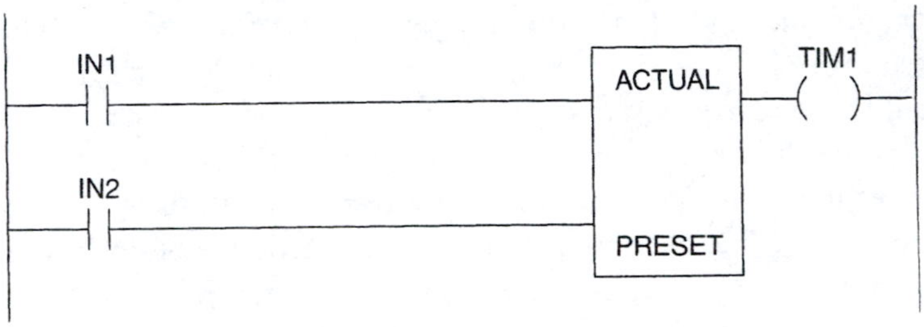

In the case of the timer, the top line is the time line. As long as the reset line is false and the time line is true, the timer will increment at the clock rate toward the PRESET value. As with the non-retentive timer, when the ACTUAL value is equal to the PRESET value, the coil associated with the timer will be energized and timer incrementing will cease. As with all timers, the PRESET value must be chosen so that the ACTUAL value will increment to the PRESET value in the time desired depending upon the clock rate.

When the RESET line of a retentive timer is true, the timer resets no matter what logical condition appears on the time line. This is called an **overriding reset.**

In some cases, the PLC manufacturer will, as with the counter, design the timer to decrement the ACTUAL value from the PRESET value toward zero with the coil associated with the timer being energized when the ACTUAL value reaches zero.

As can be seen from the explanation for timers, counters, and sequencers, these functions are very similar in operation. Frequently, the maximum number of timers, counters, and sequencers a PLC supports is represented as the total combined number. For example, a system may specify a maximum total of 64 timers/counters/sequencers. Therefore, if the program contains 20 timers, it can only contain 44 counters and/or sequencers (20 + 44 = 64 timers/counters/sequencers). The numbering of the timers, counters, and sequencers is handled differently by different manufacturers. In some cases, they are numbered sequentially (TIM1 through TIMn, CTR1 through CTRn, SEQ1 through SEQn), while in other cases they may not be allowed to share the same number (e.g., if TIM1 is present, there cannot be a CTR1 or SEQ1). Numbering and operation are dependent upon the manufacturer, and in some instances, on the model of the PLC.

A potential problem with timers can arise with some PLCs. Depending on the way the manufacturer has programmed the firmware, the timer can respond differently. Most PLCs advance timers and evaluate their state in the rung in which the timer coil appears. Some others clock the timers during I/O update. In most cases, this is an insignificant difference, but consider the program in Figure 4–16. This is a counting program that advances a counter CTR1 at regular intervals determined by the preset of TIM1. Rung 1 of this program contains a timer that clocks itself when enabled by IN1. The N/C TIM1 contact keeps the timer running until it times out. At that time, N/C TIM1 opens, resets the timer, and the cycle begins again. In most PLCs, the timer will time out on one scan and reset itself on the next, which would make all occurrences of TIM1 contacts be ON for one scan. For our program, this would advance the counter function in the second rung since the N/O TIM1 contact would apply a one scan pulse to the count input. However, in PLCs that update the timer contents during I/O update, when the timer TIM1 times out, it would switch ON during I/O update and then switch OFF in rung 1 of the very next scan. The result would be that rung 2 of the program would never "see" TIM1 switch ON, nor would any other rungs that we appended to this program. The counter CTR1 would never advance.

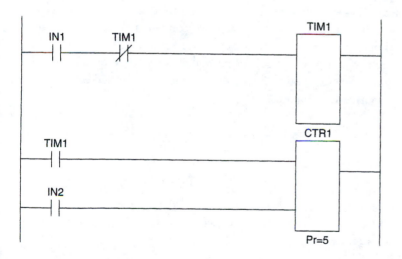

FIGURE 4–16

Timer Used as a Clock

This type of problem can be avoided by moving rung 1 to the end of the program, as shown in Figure 4–17. In this case, even if the PLC sets the timer to the ON state during I/O update, it will still scan the entire program with TIM1 ON before resetting the timer in the last rung.

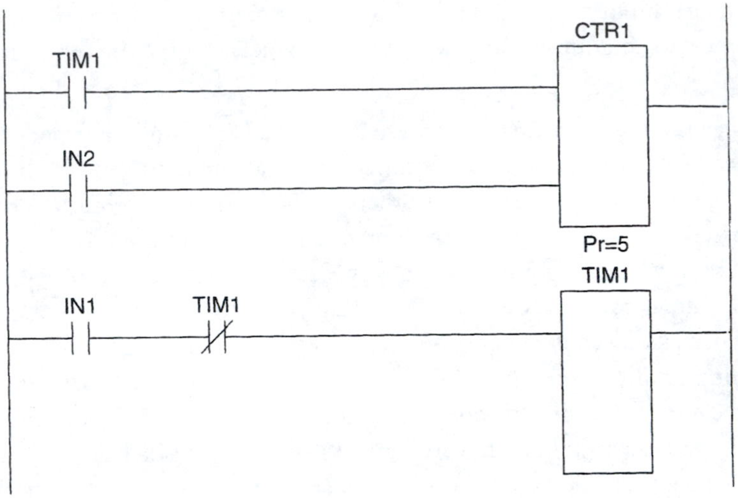

FIGURE 4–17
*Corrected Version
of Timer Used as
Clock*

Cascading Timers

If we wish to sequence events that occur at evenly spaced time intervals, we would usually construct a ladder logic flasher, which will be discussed later (or use the PLC's built-in flasher, if available), and use it to advance a sequencer. However, if each of the sequenced events must have a different on time (called a **dwell time**), a different approach is required. It is sometimes convenient to cascade non-retentive TON timers in order to create a timed sequence of events in which each event has a different dwell time. This is done by having one timer activate the next, which activates the next, and so forth. When timers are connected this way, they are said to be **cascaded timers.**

Example 4–1
Problem:

Design a ladder network that will provide a delayed-on sequence TIM1, TIM2, and TIM3. The sequence will be initiated by IN1. In operation, when IN1 is activated, TIM1 will switch ON 3.0 seconds later. Then 1.0 second later TIM2 will switch ON, followed by TIM3 2.5 seconds afterward. All signals will be reset when IN1 is switched OFF.

Solution:

Figure 4–18 shows a ladder diagram using three cascaded timers—TIM1, TIM2, and TIM3—that is, a N/O contact of TIM1 operates TIM2, and a N/O contact of TIM2 operates TIM3. In our example, TIM1, TIM2, and TIM3 have presets of 3.0s, 1.0s, and 2.5s, respectively, but any desired times may be used. The sequence is initiated by energizing IN1, which starts timer TIM1. Three seconds later, the TIM1 coil activates, which starts TIM2 timing. One second later, the TIM2 coil activates, which starts TIM3 timing. Finally, 2.5 seconds later, the TIM3 coil activates. The timing diagram shown on the right in Figure 4–18 illustrates the timed sequence of events after IN1 is switched ON. When IN1 is switched OFF, all timers in the sequence reset, and the network is immediately ready to start again.

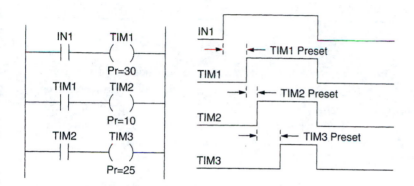

FIGURE 4–18

Cascaded Timers and Timing Diagram

It is also possible to create a similar arrangement using non-retentive TOF timers in which a sequence of OFF events is started by a single contact opening. This type of sequencer could be used to power down a machine in which a specific power down sequence is required. For example, assume a machine is hydraulically operated and that there is a fan used to cool the hydraulic fluid. In this case, we could have a cascaded TOF timer arrangement open a hydraulic dump valve first, then power down the hydraulic pump motor, then shut off the fan after a cooling period.

Flashers

A **flasher** is a ladder network that continuously cycles ON and OFF. The ON and OFF times can be unequal. Flashers are needed when it is desired to have a flashing lamp, or some other event, occur at regular, timed intervals. Most newer PLCs provide a firmware flasher, which is usually a special internal relay coil (sometimes called a **function**) that is dedicated to cycling ON and OFF at some predetermined interval (usually 0.5 second ON, and 0.5 second OFF). These "built-in" flashers are convenient because no programming is required to create the flasher, and as with all relays, the flasher contacts are available in N/O and N/C configurations. However, the ON and OFF times are fixed. If other ON and OFF times are desired, the programmer must construct a custom flasher network in ladder logic. This is done using two cascaded timers that are connected in a regenerative fashion; that is, the first timer enables the second, and the second timer disables the first. Flashers can either run non-stop, or they can be started and stopped by simply inserting a controlling contact in the loop.

Example 4–2
Problem:

A machine's electrically operated automatic oiler is connected to dispense oil when OUT1 of the PLC is ON. It has been determined that, because it is over-oiling the machine, we wish to reduce the amount of oil dispensed by periodically cycling it ON and OFF. Design a PLC program that will operate the oiler connected to output OUT1 when input IN1 is ON. When IN1 is ON, the output OUT1 is to cycle continuously ON for 0.5 second and OFF for 1.0 second.

Solution:

Since there are two times specified in this problem (0.5 second and 1.0 second), we will need two non-retentive timers, and we will need to connect the timers so that the first enables the second, and the second disables the first (as illustrated in Figure 4–19). In rung 1 of this network, IN1 activates TIM1. In rung 2, after TIM1 times out (0.5 second after IN1 is switched ON), TIM2 is started timing. Then 1.0 second later, when TIM2 times out, the N/C TIM2 contact in rung 1 opens which resets and restarts TIM1. Since TIM1 is restarted, this will consequently reset TIM2 in rung 2. Therefore, when IN1 is ON, TIM1 will continuously cycle OFF for 0.5 second and ON

FIGURE 4–19

Flasher

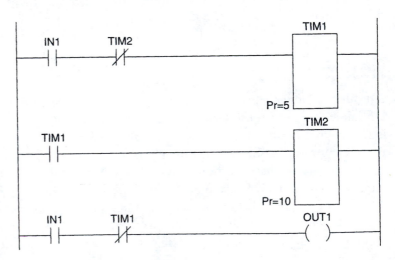

for 1.0 second. Rung 3 of our program is enabled by IN1. As long as IN1 is ON, OUT1 is simply the inverse of TIM1 (because of the N/C TIM1 contact in the rung). Since TIM1 is OFF for 0.5 second and ON for 1.0 second, OUT1 will be ON for 0.5 second and OFF for 1.0 second. (Note: One advantage of using a PLC to perform this oiler function is that, if further adjustment of the oiler output is needed, it is simply done by adjusting the preset of TIM1.)

Timed One Shot

By definition, a common one shot in PLC programming is a contact that remains ON for only one scan. However, it is also possible to construct a timed one shot in which a contact may be triggered by a contact closure and the contact will remain ON for some timed period. Figure 4–20 shows such a network. In this case, when IN1 is activated, TIM1 begins timing and OUT1 switches ON. The N/O OUT1 contact seals IN1 and keeps the network active. When the timer TIM1 times out 3.5 seconds later, the N/C TIM1 contact opens which switches OFF the network and opens the sealing contact OUT1. Therefore, OUT1 will be activated when IN1 switches ON and will remain ON for the preset of TIM1, in this case 3.5 seconds. This network assumes that IN1 will be ON for a time period less then the preset of IN1; otherwise, it will be necessary to perform a one scan, one shot operation on IN1.

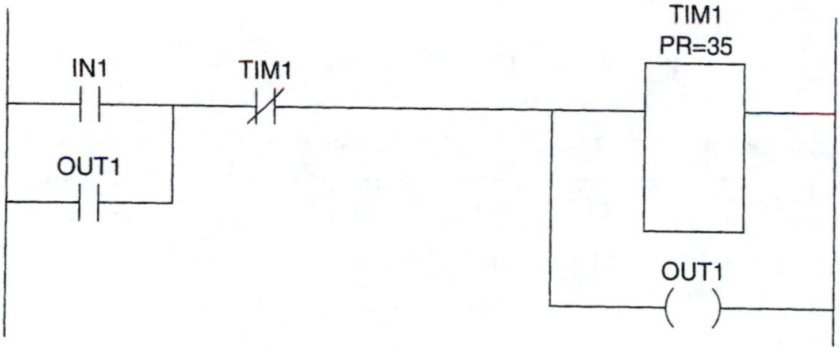

FIGURE 4–20

Timed One Shot

Notice that the network in Figure 4–20 has two coils, TIM1 and OUT1, in parallel. Some PLC manufacturers do not allow this programming style and require that there be only one coil per rung. If this is the case, the network can be broken into the two-rung one shot shown in Figure 4–21, with identical contact arrangements on the left and each of the two coils TIM1 and OUT1 on the right. If this is done, it is important that the rung containing the coil OUT1 appear first in the program. Otherwise, the N/C OUT1 in the TIM1 rung will never switch off, which will cause the one shot to cycle continuously.

FIGURE 4–21
Two-Rung Timed One Shot

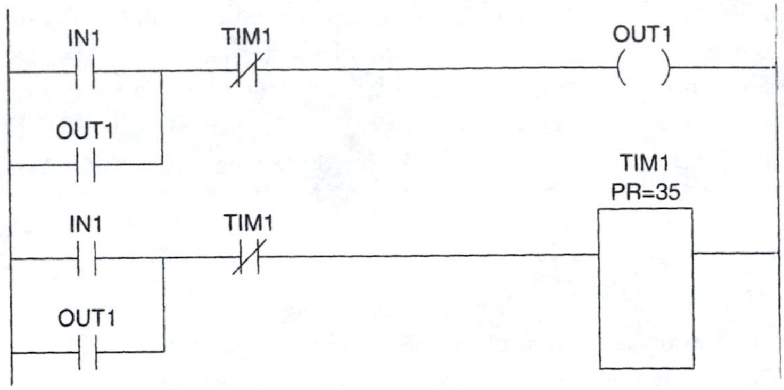

Timed Sequencer

A flasher and sequencer can be utilized to perform an operation called the timed sequencer. This is a function commonly used when a series of events must occur at predetermined intervals and each interval is an integer multiple of some basic unit of time. For example, we could have a washing machine controlled by a timed sequencer that has a basic time interval of one minute. This means that the shortest event that can be controlled will be one minute, but we could also have longer events as long as they are integer multiples of one minute. In some cases, we draw timing charts for these devices, which show the sequence of events that the machine will execute.

Figure 4–22 shows a timing chart for a hypothetical washing machine. The machine has four basic functions—fill, agitate, spin, and pump. Using combinations of the four functions, we can program the machine to perform logical operations to wash and rinse/spin dry our clothes. It should be noted that our washing machine example is not very practical, mainly because the fill operation is timed. Since water pressure can vary, the fill operation should instead be controlled by a float switch. However, we will time the fill operation for this illustration in order to make the example less complex.

FIGURE 4–22
Washing Machine Timing Chart

Minute	1	2	3	4	5	6	7	8	9	10	11	12	13	14	15	16	17	18	19	20	21	22
	Wash										Rinse/Spin Dry											
Fill	■	■							■		■	■							■			
Agitate			■	■	■	■	■						■	■	■	■	■					
Spin									■	■									■	■	■	■
Pump								■	■	■								■	■	■	■	■

From the timing chart, we can see that the wash cycle consists of a two-minute fill, five minutes of agitate, and three minutes to pump out the water, the final two minutes of which is the spin function. Also during spinning, we open the fill valve for one minute to help rinse the suds from the clothes. The rinse/spin dry cycle is the same except that the spin function has been extended two extra minutes to remove more of the water from the clothes.

The ladder diagram of the PLC program to implement the washing machine controller is shown in Figure 4–23. As we discuss each rung of the program, follow along in the figure.

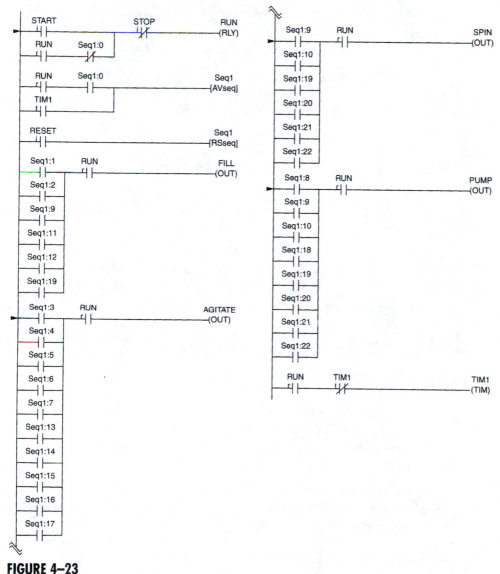

FIGURE 4–23

Washing Machine Program

Rung 1 is a standard latched starter network. RUN is switched ON by external input START and switched OFF by either external input STOP or when the sequencer reaches count zero (i.e., it completes a cycle).

Rung 2 clocks the sequencer. The sequencer will be advanced from zero to one when RUN is activated and at all other times by the clock, TIM1.

Rung 3 uses external input RESET to reset the sequencer to zero.

Rung 4 controls output FILL (the water fill valve). As indicated in the timing chart in Figure 4–22, FILL is activated when RUN is active and when the sequencer is at counts 1, 2, 9, 11, 12, and 19.

Rung 5 controls output AGITATE. As indicated in the timing chart in Figure 4–22, AGITATE is activated when RUN is active and when the sequencer is at counts 3 through 7 and 13 through 17.

Rung 6 controls output SPIN. As indicated in the timing chart in Figure 4–22, SPIN is activated when RUN is active and when the sequencer is at counts 9, 10, and 19 through 22.

Rung 7 controls output PUMP. As shown in the timing chart in Figure 4–22, PUMP is activated when RUN is active and when the sequencer is at counts 8 through 10 and 18 through 22.

Rung 8 is the clock rung. When RUN is active, the one-minute clock, TIM1, continuously cycles and provides the program with one scan pulses at one-minute intervals. Note that this rung has been put at the end of the program to avoid any clocking problems.

4-11 Master Control Relays and Control Zones

In an earlier chapter, the concept of control zones, or master control relays, was discussed. These are used in cases where the designer wishes to disable or enable entire sections of the control circuit. Zones, or MCRs, can also be implemented in ladder logic.

Ladder logic control zones are designated by two zone coils, **zone begin** and **zone end** (or Interlock On and Interlock Off, or MCR On and MCR Off). The zone begin is a coil that is operated by a contact that the programmer wishes to use as the controlling function (in our example, this would be the input that is connected to the MAINTENANCE switch). The zone-begin rung is followed by all of the rungs that we wish to have enabled or disabled by the control. These rungs are then followed by a zone-end rung. The zone-end function is a coil that is generally in a rung by itself (i.e., there are no contacts controlling the coil). It is simply used as a marker to designate the end of the zone.

In operation, as the PLC is scanning the program, when it encounters a zone-begin rung, it evaluates the rung to determine if it is ON or OFF. If the rung is ON, then the PLC continues scanning with the very next rung. All rungs within the zone are evaluated normally. However, if the zone-begin rung is evaluated as OFF, the PLC skips all the rungs within the zone. Normal scanning resumes with the rung that immediately follows the zone-end rung.

Figure 4–24 shows an example of a control zone in a segment of a PLC program. In this case, the first and last rungs of the program segment are evaluated by the PLC on every scan. However, the rungs containing OUT101 and OUT113 are within the zone and are thereby controlled by the zone-begin rung. Therefore, if CR1 is ON, the rungs containing OUT101 and OUT113 are evaluated normally; however, if CR1 is OFF, the zone is disabled and the rungs containing OUT101 and OUT113 are not evaluated.

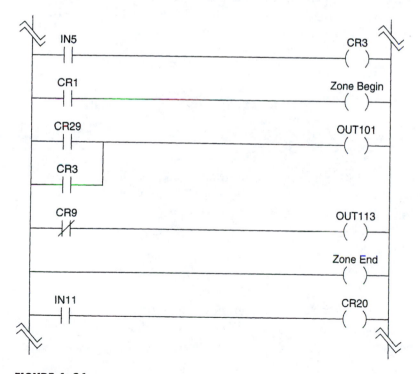

FIGURE 4–24
Program Segment Illustrating Zone Control

The exact names of the zone-begin and zone-end coils vary widely between PLC manufacturers, as does the program structure. Also, when a zone is disabled, some manufacturers choose to turn off all coils within the zone, while others leave the coils in their last known state. Therefore, it is imperative that the programmer study the programming manual for the particular PLC being used to assure that the zone commands are completely understood prior to using them.

Summary

As we have seen in this chapter, advanced PLC programming techniques allow the system designer more power and flexibility than can be practically produced by hard wired relay logic. By carefully arranging the ordering of rungs of logic, the programmer has the capability to construct both static logic functions and active logical devices such as oscillators, flip flops, pulsers, and one shots. Also, by utilizing the more advanced features available in PLC ladder logic language such as the timer, counter, sequencer, and control zones, the programmer has the capability to have the PLC perform extremely complex control functions. Altogether, these advanced programming techniques make the PLC the obvious choice over wired relay logic for any complex control system.

Review Questions

1. Draw the ladder rung for an R-S Flip Flop that will energize when both IN1 AND IN2 are ON and will de-energize when both IN3 AND IN4 are ON. The condition where all inputs are ON will not be a defined state for this problem (i.e., it will not be allowed to occur so you do not have to plan for it).

2. Draw the ladder diagram for a T Flip Flop CR1, which will toggle only when IN1 and IN2 are both OFF. The clock is IN3.

3. Design a flip flop ladder network that will convert a N/O momentary pushbutton switch connected to IN1 to a maintained operation that turns ON output OUT1. When the button is pressed, OUT1 turns ON. When the button is released and pressed again, OUT1 turns OFF.

4. Design a ladder program that will provide an internal relay CR1 that turns ON for three seconds when the program begins execution. Afterward, it turns OFF and remains OFF until the program is restarted.

5. Design a ladder network for a "pulse stretcher." When IN4 switches ON, OUT3 energizes for 1.5 seconds, then

switches OFF. The ON time of IN4 should have no effect on the ON time of OUT3 (i.e., IN4 could be ON for less than or more than 1.5 seconds without changing the OUT3 ON time).

6. Develop the ladder program for a system of two T Flip Flops that will function as a two-bit binary counter. The least significant bit should be CR1, and the most significant bit should be CR2. The clock input should be IN17.

7. Develop the ladder diagram for a three-bit shift register using J-K Flip Flops that will shift each time IN1 is switched from OFF to ON. The input for the shift register is to be IN2. The three coils for the shift register may have any coil numbers you choose.

8. Design the ladder diagram for a device that will count parts as they pass by an inspection stand. The sensing device for the PLC is a switch that will close each time a part passes. This sensing switch, named SENSOR, is connected to IN1 of the PLC. A reset switch, IN2, named C-RESET is also connected to the PLC to allow the operator to manually reset the counter. After fifteen parts have passed the inspection stand, the PLC is to reset the counter to begin counting parts again and turn on a light that must stay on until reset by a second reset switch, L-RESET, connected to IN3. The output from the PLC that lights the light is OUT111.

9. Design the ladder diagram for a program needing a timer that will cause coil CR24 to energize for one scan every 5.5 seconds.

Programming Projects

1. Write a ladder logic program as defined here. Assign the following inputs and outputs in your program:

IN1 named START
IN2 named STOP
IN3 named FORM
IN4 named FAIL
OUT1 named MOTOR
OUT2 named EJECT
OUT3 named CYCLE

All inputs to the PLC are N/O momentary pushbuttons. Although this program can be completed without internal relays, if you need internal relays, you may name them as desired.

Functions:

a. When the program starts, all outputs are to be OFF.

b. When START is pressed, MOTOR should switch ON. MOTOR should stay on even after START is released. MOTOR is switched OFF by pressing STOP.

c. If FORM is pressed and MOTOR is ON, CYCLE switches ON. CYCLE stays on only during the time that FORM is being pressed. CYCLE cannot operate if MOTOR is OFF.

d. If FAIL is pressed and MOTOR is OFF, EJECT will switch ON. Even after FAIL is released, EJECT will stay ON. EJECT automatically switches OFF when MOTOR is again switched ON. EJECT cannot operate while MOTOR is ON.

2. The entrance/exit gates at a theme park have turnstiles that can produce a set of logical signals each time someone passes through the gate. It is desired to design a PLC program that will indicate both the total number of people that have attended the park and the present number of people in the park. A master RESET button should be provided that will allow both counters to be cleared at the beginning of each day. Design a PLC program that will track the activity for one turnstile. There will be three inputs—PHASEA, PHASEB, and RESET. There will be no outputs. Your program should be able to track at least 999 total guests. The turnstile contains two switches, PHASEA and PHASEB, that are mounted so that as a guest passes through the turnstile the two switches will turn on in a sequence. The sequence depends on whether the guest is entering or exiting the park. When someone enters the gate, switch PHASEA outputs a signal before switch PHASEB. When someone exits the gate, switch PHASEB outputs a signal before PHASEA. However, since the two switches are located close together, there will always be some overlap in their pulse outputs unless the guest starts

through a gate and then backs out. If the guest backs out, it will create a false trigger that should not be counted. The timing diagram in Figure 4–25 shows how this will work.

Notice that false triggers should not cause the program to count because PHASEA and PHASEB do not overlap. Two counters will be needed. One will count up only and indicate the total number of people that have attended the park for the day. The second counter counts both up (when someone enters) and down (when someone exits) and indicates the number of people presently in the park.

3. One of the water rides at a water theme park is the water slide racer that is based on a drag racing theme. On this ride, five guests sit on mats and slide down individual, equal-length water slides side-by-side in a race to the finish. A PLC connected to this ride is to perform two basic functions:

   a. A set of starting lights similar to those at a drag racing track (called "Christmas Tree Lights") signal when to start. The light combination consists of one red light, three yellow lights, and one green light arranged vertically with red on top. Normally,

FIGURE 4–25

Turnstile Timing Diagram

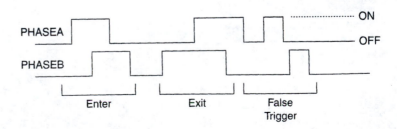

the RED light is ON. However, when the ride operator presses a momentary START button, the other lights sequence ON at one-second intervals. The sequence is YELLOW1, YELLOW2, YELLOW3, GREEN. The GREEN remains ON for 10 seconds to allow the racers to complete their run. Then the GREEN goes OFF and the RED illuminates to indicate that it is time for the next racers to take their mark. Only one light is ON at any time. Holding the START button down for a long period of time should not prevent the lights from completing their cycle and should not cause another cycle to start (i.e., it should have an anti-repeat feature).

b. An optical sensor is embedded in the raceway at the finish line of each slide. When racers pass over their respective sensor, daylight is blocked from the sensor, which, in turn, switches on a PLC input. The sensors are named LANE1 through LANE5. There is one winner lamp for each lane. The PLC determines which lane is the winner and lights the winner lamp for that particular lane. Remember, the first racer to the finish line gets the winner lamp—the other racers do not. In addition, even though the winning racer will pass over the sensor only momentarily, the winner light should remain ON until the RED lamp at the top of the hill is lit.

The inputs to the PLC are as follows:

START = the operator's start button, a momentary pushbutton

LANE1 = the finish line sensor for lane 1

LANE2 = the finish line sensor for lane 2

LANE3 = the finish line sensor for lane 3

LANE4 = the finish line sensor for lane 4

LANE5 = the finish line sensor for lane 5

The PLC outputs are as follows:

RED = the top starting light

YELLOW1 = the second starting light

YELLOW2 = the third starting light

YELLOW3 = the fourth starting light

GREEN = the bottom starting light (that signals for the racers to start)

WIN1 = the winner light for lane 1

WIN2 = the winner light for lane 2

WIN3 = the winner light for lane 3

WIN4 = the winner light for lane 4

WIN5 = the winner light for lane 5

A typical cycle should happen as follows:

1. Initialize with RED light ON, all other lights OFF, wait for START button

2. START button pressed, RED immediately turns OFF, YELLOW1 lights for one second

3. YELLOW1 turns OFF, YELLOW2 lights for one second

4. YELLOW2 turns OFF, YELLOW3 lights for one second

5. YELLOW3 turns OFF, GREEN lights, ten-second timer starts

6. Racers cross the finish line, winning lane is determined, and appropriate winner light (WIN1 through WIN5) switches ON.

7. When ten-second timer times out, turn off winner light and GREEN light, turn on RED light, go back to step 1.

Your program should be able to handle unusual situations gracefully. In particular, it should not "hang" if it is started with no racers on the slides (i.e., no winner), and it should not do unusual things if the START button is pressed multiple times during the same cycle. In case of a tie, the lower numbered lane wins.

4. A PLC controls five sets of stage lights for an auditorium. For one particular stage production, the various acts and scenes require six different lighting sequences, each involving a different combination of the five light sets. The stagehand in charge of lighting has a control panel that controls the PLC. There are two switches on the panel:

STEP—a momentary N/O pushbutton switch that makes the PLC step to the next lighting sequence.

UPDN—a maintained switch that makes the sequence advance when ON and reverse when OFF (reverse is used for repeating sequences during rehearsal).

The lighting sequence required is shown in Figure 4–26.

For this problem, there will be two PLC inputs, STEP and UPDN, and five PLC outputs, LIGHTS1 through LIGHTS5. For the purpose of special effects, LIGHTS5 should flash continuously for 0.3 second ON and 0.3 second OFF when activated. All the lights should operate in the sequence shown in the table and should sequence both forward and reverse. The sequencer should be circular (i.e., when sequencing forward, step 0 follows step 6, and when sequencing in reverse, step 6 follows step 0).

FIGURE 4–26

Sequence Table for Stage Light Problem

	STEP 0	STEP 1	STEP 2	STEP 3	STEP 4	STEP 5	STEP 6
LIGHTS1		■	■			■	
LIGHTS2		■			■	■	
LIGHTS3			■		■	■	■
LIGHTS4		■				■	
LIGHTS5		FLASH	FLASH		FLASH	FLASH	

Mnemonic Programming Code

Objectives

Upon completion of this chapter, you will know:

- why mnemonic code is used in some cases instead of graphical ladder language.

- some of the more commonly used mnemonic codes for AND, OR, and INVERT operations.

- how to represent ladder branches in mnemonic code.

- how to use stack operations when entering mnemonic coded programs.

Introduction

All discussions in previous sections have considered only the ladder diagram in program example development. However, now we will consider how the ladder diagram is entered into the programmable logic controller. In higher-order controllers, this can be accomplished through the use of dedicated personal computer software that allows the programmer to enter the ladder diagram as drawn. The software then takes care of translating the ladder diagram into the code required by the controller. In the more basic controllers, this translation has to be performed by the programmer and entered by hand into the controller. This type of language and the procedure for translating the ladder diagram into the required code will be discussed in this chapter. This will be accomplished by retracing some of the examples and ladder diagrams developed in earlier chapters and translating them into the mnemonic code required to program a general controller. This controller will be programmed in a somewhat generic type of code. As we progress, comparisons will be presented with similar types of statements found in various other controllers.

To develop a program for one of these other controllers, the reader will have only to adapt to the syntax required by that type.

Although mnemonic PLC programming language is somewhat more difficult to comprehend than graphical ladder logic, it is necessary for the system designer to be familiar with the language since it is used by many handheld PLC programmers. These programmers are small, lightweight, and convenient to use when performing minor troubleshooting and simple program modifications. By using the handheld programmer, we avoid the need to carry a more expensive and cumbersome laptop into a manufacturing environment.

5-1 AND Ladder Rung

We will begin with the ladder diagram of Figure 5–1. This is the AND combination of two contacts, IN1 and IN2, controlling coil OUT1.

FIGURE 5–1

Ladder Diagram for AND Function

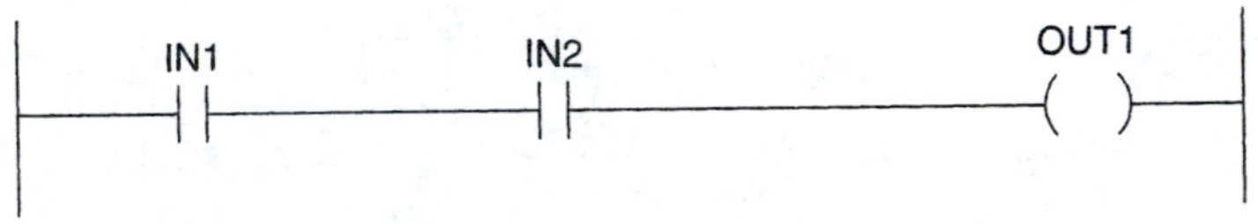

Ladder diagrams are made up of interconnected branches of contact logic that control a coil. For instance, IN1 AND IN2 can be considered a branch. Although this rung has only one branch, we will see examples of multiple branches later. The code command that alerts the controller to the beginning of a branch is LD. The **LD command** tells the controller that the following set of contacts composes one branch of logic. The complete contact command code for our branch in Figure 5–1 is:

LD   IN1

AND   IN2

These lines tell the controller to start a branch with IN1 and AND it with contact IN2. LD commands are terminated with either another LD command or a coil command. In this case, a coil command would terminate because there are no more contacts contained in the ladder. The coil command is STO. The complete contact and coil commands for this rung of logic are:

LD   IN1

AND   IN2

STO   OUT1

The STO command tells the controller that the previous logic is to control the coil that follows the STO command. Each line of code must be input into the controller as an individual command. The termination command for a line of code is generally ENTER.

NOTE: Two types of terminators have been described and should not be confused with each other. Commands are terminated by another command (a software item), while lines are terminated with ENTER (a hardware keyboard key).

The complete command listing for this ladder rung, including termination commands, is:

```
LD      IN1     ENTER
AND     IN2     ENTER
STO     OUT1    ENTER
```

The commands may be entered using a hand-held programmer, dedicated desktop programmer, or a computer containing software that will allow it to operate as a programming device. Each controller command line contains (1) a command, (2) the object (or operand) of the command, and (3) a terminator (the ENTER key). In the case of the first line, LD is the command, IN1 is the object of the command, and the ENTER key is the terminator. Each line of code will typically consume one word of memory, although some of the more complex commands will consume more than one word. Examples of commands that may consume more than one word of memory are math functions, timers, and counters.

5-2 Entering Normally Closed Contacts

Notice that the rung of Figure 5–1 has only normally open contacts and no normally closed contacts. We will now investigate how the command lines would change with the inclusion of a normally closed contact in the rung, as illustrated in Figure 5–2. Notice that normally open contact IN1 has now been replaced with normally closed contact IN1.

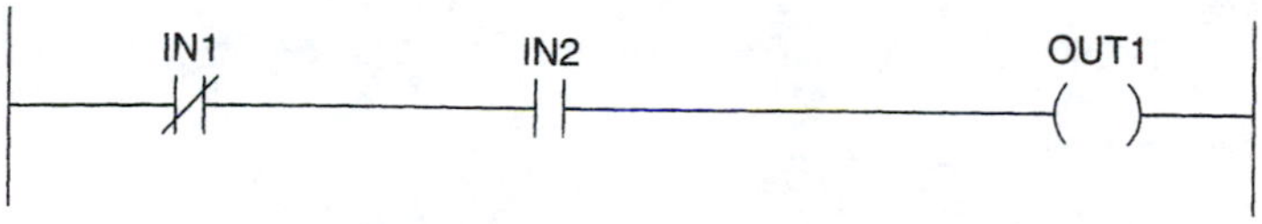

FIGURE 5–2
Rung With a Normally Closed Contact

To indicate a normally closed contact to the PLC, the term NOT is associated with the contact number. This may take different forms in different controllers depending on the program method used by the manufacturer. Using the same form as in the previous example, the command lines for this rung would appear as follows:

LD	NOT IN1	ENTER
AND	IN2	ENTER
STO	OUT1	ENTER

As stated previously, different PLCs may use different commands to perform some functions. For instance, the Mitsubishi PLC uses the command LDI (LD INVERSE) instead of LD NOT. This requires a single keystroke instead of two keystrokes to input the same command.

If the normally closed contact had been IN2 instead of IN1, the command lines would have to be modified as follows:

LD	IN1	ENTER
AND	NOT IN2	ENTER
STO	OUT1	ENTER

If using the Mitsubishi PLC, the AND NOT command would be replaced with the ANI (AND INVERSE) command.

5-3 OR Ladder Rung

Now, let us translate the ladder rung of Figure 5–3 into machine code.

FIGURE 5–3

Ladder Diagram for OR Function

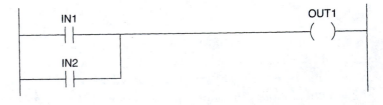

As can be seen, the contact logic for Figure 5–3 is an OR connection controlling coil OUT1. Following the same steps as Figure 5–1, the command lines for this rung are:

LD	IN1	ENTER
OR	IN2	ENTER
STO	OUT1	ENTER

Notice again each line contains a command, an object, and a terminator. In the case of this particular controller, the coil has a descriptive label (OUT) associated with it to tell us that this coil is an output for the controller. In some controllers, this is not the case. Some controllers designate the coil as output or internal depending upon the number assigned to it. For instance, output coils may be coils having numbers between 100 and 110 with internal coils having numbers between 111 and 200. Systems that are composed of plug-in modules may set the output coil and input contact numbers by the physical location of the modules in the system.

5-4 Simple Branches

Now consider the ladder diagram of Figure 5–4, which is a more complicated AND-OR-AND logic containing two branches.

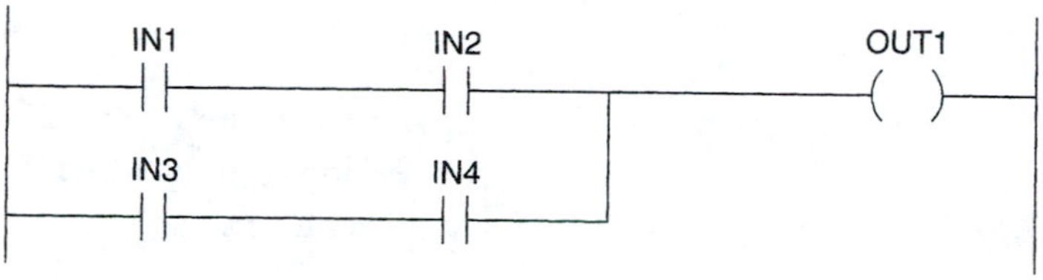

FIGURE 5–4

Ladder Diagram AND - OR - AND Function

The previous examples have had only a single branch in each case. A branch may be defined as a single logic expression contained in the overall Boolean expression for a rung. In the case of Figure 5–4, the Boolean expression is shown in Equation 5–1.

$$OUT1 = (IN1)(IN2) + (IN3)(IN4) \qquad (5\text{--}1)$$

As can be seen in the Boolean equation, there are two logic expressions OR'ed together; IN1 AND IN2 OR'ed with IN3 AND IN4. Each of the two expressions is called a branch of logic and must be handled separately when being input into the controller. Incorrectly entering the branches will result in improper (and possibly dangerous) operation of the PLC. In cases where entering a branch incorrectly violates the internal controller logic, an error message will be generated and the entry disallowed. In cases where the entry does not violate controller logic, operation will be allowed and could cause bodily injury to personnel if the controller is in an application where it operates dangerous machinery. The configuration of the logic in Figure 5–4 utilizes four contacts, IN1, IN2, IN3, and IN4, controlling an output coil OUT1. As discussed in earlier chapters, coil OUT1 will energize when (IN1 AND IN2) OR (IN3 AND IN4) is true. The first line (IN1 AND IN2) is entered in the same manner as described in the previous examples: (Note: the 1. and 2. are line numbers for our reference only and are not entered into the PLC)

1. LD   IN1   ENTER

2. AND   IN2   ENTER

For the next line (IN3 AND IN4), we must start a new branch. This is accomplished through the use of another LD statement and a portion of PLC memory called the stack.

As program commands for a rung are entered into the PLC they go into what we will call an active memory area. The stack is an additional memory area set aside for temporarily storing portions of the commands being input. Each time an LD command is input, the controller transfers all logic currently in the active area to the stack. When the first LD command of the rung is input, there is nothing in the active area to transfer to the stack. The next two lines of code would be as follows:

3. LD   IN3   ENTER

4. AND   IN4   ENTER

The LD command for IN3 causes the previous two lines of code (lines 1 and 2) to be transferred to the stack. After lines 3 and 4 have been input, lines 1 and 2 will be in the stack, and lines 3 and 4 will be in the active memory area.

The two areas, active and the stack, may now be OR'ed with each other using the command OR LD. This tells the controller to retrieve the commands from the stack and OR that code with what is in the active area. *The resulting expression is left in the active area.* This line of code would be input as shown here:

5. OR   LD   ENTER

Up to now, most controllers would have recognized the same type of commands (AND, OR). The LD and OR LD commands will, however, appear differently in various controllers depending upon how the manufacturer wishes to design the controller. For instance, the Allen Bradley SLC-100 handles branches using **branch open** and **branch close** commands instead of LD and OR LD. Mitsubishi controllers use OR BLK instead of OR LD. Some controllers will use STR in place of LD. Also, in some cases, all coils may be entered as OUT with an associated number that specifies it as an output or internal coil. The concept is generally the same, but commands vary with controller manufacturer type and even by model in some units.

Adding the coil command for the ladder diagram of Figure 5–4, the commands required for the entire rung are as follows:

1. LD   IN1   ENTER

2. AND   IN2   ENTER

3. LD   IN3   ENTER

4. AND   IN4   ENTER

5. OR   LD   ENTER

6. STO   OUT1   ENTER

As a matter of comparison, if the previous program had been input into a controller that utilizes STR instead of LD, and OUT instead of STO, the commands would be:

STR   IN1   ENTER

AND   IN2   ENTER

STR   IN3   ENTER

AND   IN4   ENTER

OR   STR   ENTER

OUT   101   ENTER

The 101 associated with the OUT statement designates the coil as number 101, which, in the case of this particular PLC would, by design, cause it to be an internal or output coil as defined by the PLC manufacturer. As can be seen, the structure is the same, but the commands vary as required by the PLC brand being used.

Let us now discuss the ladder rung of Figure 5–5. Recall that this is an OR-AND-OR logic rung having the Boolean expression shown in Equation 5–2.

$$OUT1 = (IN1 + IN2)(IN3 + IN4) \qquad (5\text{–}2)$$

Two branches can be seen in this expression: (IN1 OR IN2) and (IN3 OR IN4). These two branches are to be AND'ed together.

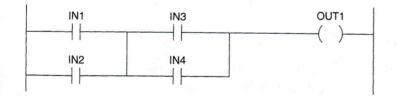

FIGURE 5–5

Ladder Diagram to Implement OR-AND-OR Function

Using the LD and STO commands and the same approach as in the previous examples, the command structure would be as follows:

1. LD   IN1   ENTER

2. OR   IN2   ENTER

3. LD   IN3   ENTER

4. OR   IN4   ENTER

5. AND   LD   ENTER

6. STO   OUT1   ENTER

As in the previous example, the LD command in line 3 causes lines 1 and 2 to be transferred to the stack. Line 5 causes the active area and stack to be AND'ed with each other.

5-5 Complex Branches

Now that we have discussed basic AND and OR techniques and simple branches, we will examine a rung with a more complex logic expression to illustrate how multiple branches would be programmed. Consider the rung of Figure 5–6. Notice that there are multiple logic expressions contained in the overall ladder rung. The Boolean expression for this rung is shown in Equation 5–3.

FIGURE 5–6

Complex Ladder Rung

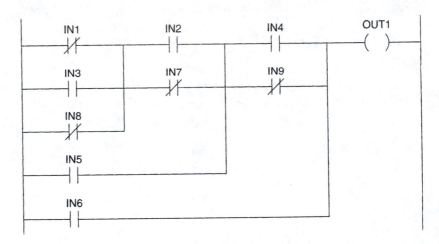

$$OUT1 = \{[(\overline{IN1} + IN3 + \overline{IN8})(IN2 + \overline{IN7})] + IN5\} \{IN4 + \overline{IN9}\} + IN6 \qquad (5\text{–}3)$$

To develop the program commands for this logic, we will begin with the innermost logic expression. This is IN1′ + IN3 + IN8′. The commands to enter this expression are:

1. LD   NOT IN1   ENTER

2. OR   IN3   ENTER

3. OR   NOT IN8   ENTER

The next expression to enter is IN2 + IN7', which must be AND'ed with the expression now in the active area. To accomplish this, the logic in the active area must be transferred to the stack, and then the next expression entered and AND'ed with the stack. The command lines for this are:

4. LD IN2   ENTER

5. OR NOT IN7 ENTER

6. AND LD   ENTER

Line 4 places the previous expression in the stack and begins the new expression, and line 5 completes the new expression. Line 6 causes the stack logic to be retrieved and AND'ed with the active area with the result left in the active area. Referring to Equation 5–3, notice that the expression located in the active area must now be OR'ed with IN5. This is a simple OR command since the first part of the OR is already in the active area. The command to accomplish this is:

7. OR IN5 ENTER

Now the active area contains [(IN1' + IN3 + IN8') (IN2 + IN7')] + IN5. This must be AND'ed with {IN4 + IN9'}. To input this expression, we must transfer the logic in the active area to the stack, input the new expression, retrieve the stack, and AND it with the expression in the active area. These commands are:

8. LD IN4   ENTER

9. OR NOT IN9 ENTER

10. AND LD   ENTER

Line 8 transfers the previous expression to the stack and begins input of the new expression, line 9 completes entry of the new expression, and line 10 AND's the stack with the new expression. The active area now contains {[(IN1' + IN3 + IN8') (IN2 + IN7')] + IN5} {IN4 + IN9'}. As may be seen in Equation 5–3, all that remains is to OR this expression with IN6 and add the OUT1 coil command. The command lines for this are:

11. OR IN6 ENTER

12. STO OUT1 ENTER

Combining all the command lines, we have the program here:

1. LD NOT IN1 ENTER

2. OR IN3   ENTER

3. OR NOT IN8 ENTER

4. LD IN2   ENTER

5. OR	NOT IN7	ENTER
6. AND	LD	ENTER
7. OR	IN5	ENTER
8. LD	IN4	ENTER
9. OR	NOT IN9	ENTER
10. AND	LD	ENTER
11. OR	IN6	ENTER
12. STO	OUT1	ENTER

The previous example should be reviewed to be sure operation involving the stack is thoroughly understood.

Summary

It is most likely that the system designer will need to use the mnemonic PLC programming language after installation of a system; that is, when doing minor debugging, simple program updates, troubleshooting, or fine tuning of a machine's control parameters. That is because it is easier and faster to connect a handheld programmer to a PLC once it is in a manufacturing environment (rather than use a desktop or laptop computer), and most handheld programmers use the mnemonic PLC language. Although this may seem unwieldy to a newcomer, by gaining an understanding of the common mnemonic structure explored in this chapter, is it relatively easy to adapt to any nuances in a PLC manufacturer's mnemonic language.

Review Questions

1. Draw the ladder diagram, and write the mnemonic PLC code for a program that will accept inputs from switches IN1, IN2, IN3, IN4, and IN5 and will energize coil OUT123 when one and only one of the inputs is ON.

2. Draw the ladder diagram, and write the mnemonic PLC code for an oscillator named CR3.

3. Write the mnemonic PLC code for the ladder diagram of Figure 5–7.

FIGURE 5–7

Ladder Diagram for Problem 3

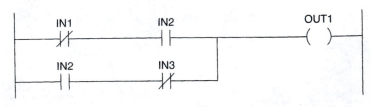

4. Draw the ladder diagram, assign contact and coil numbers, and write the mnemonic PLC code for the J-K Flip Flop.

5. Draw the ladder diagram, assign contact and coil numbers, and write the mnemonic PLC code for a T Flip Flop.

6. Write the mnemonic PLC code for the ladder diagram of Figure 5–8.

7. Write the mnemonic PLC code for the ladder diagram of Figure 5–9.

8. Assume that the mnemonic PLC code to enter a standard-speed non-retentive timer coil T1 with a preset time of 5.5 seconds is

STO T1 55 ENTER

Write the mnemonic PLC code for the flasher ladder diagram of Figure 5–10.

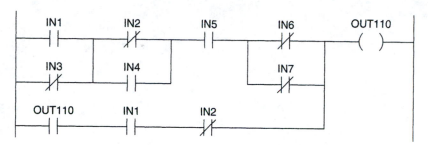

FIGURE 5–8

Ladder Diagram for Problem 6

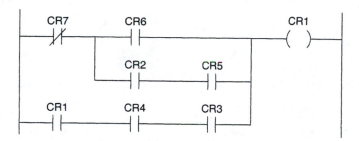

FIGURE 5–9

Ladder Diagram for Problem 7

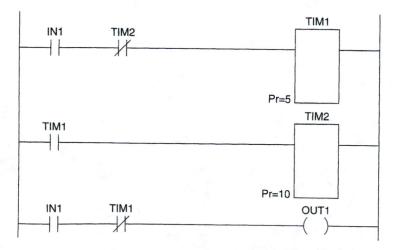

FIGURE 5–10

Ladder Diagram for Problem 8

9. Write the mnemonic PLC code for the Boolean expression:

$$OUT100 = \{(\overline{IN3} + OUT115 + (IN7)(IN11)\} + \overline{OUT101}$$

10. Write the mnemonic PLC code for the Boolean expression:

$$OUT110 = \overline{OUT105} + \{(IN30 + OUT110)(\overline{IN22}) + IN11\}$$

Wiring Techniques

CHAPTER

6

Objectives

Upon completion of this chapter, you will know:

- how to provide ac or dc power to a PLC.
- various types of PLC input configurations.
- how to select the best PLC input configuration for an application.
- how to connect external components to PLC inputs.
- various types of PLC output configurations.
- how to select the best PLC output configuration for an application.
- how to connect PLC outputs to external components.

Introduction

A very important subject often overlooked in the study of programmable controllers is how to connect the PLC to the system being controlled. This involves connections of such devices as limit switches, proximity detectors, photoelectric detectors, external high current contactors and motor starters, lights, and a vast array of other devices that can be utilized with the PLC to control or monitor systems. Wiring a device to the PLC involves the provision of proper power to the devices, sizing of wiring to ensure current-carrying capacity, routing of wiring for safety and to minimize interference, ensuring that all connections are made properly and to the correct terminals, and providing adequate fusing to protect the system.

PLC systems typically involve the handling of circuitry operating at several different voltage and current levels. Power to the PLC and other devices may require the connection of 120 VAC while photoelectric and

115

proximity devices may require 24 VDC. Motors being controlled by the PLC may operate at much higher voltage levels such as 240 or 480 VAC, three-phase. Current for photoelectric and proximity devices are in the range of milliamperes while motor currents are typically much higher and depend upon the size of the motor.

Except for main power, PLC connections are confined to connecting inputs to sensing devices and switches and connecting outputs to devices being controlled (lamps, motor starters, contactors). This main area of concern for the programmer is the focus of this chapter. We will also touch on the other areas as required while discussing input and output connections.

6-1 PLC Power Connection

The power requirement for the PLC being used will vary depending upon the model selected. PLCs are available that operate on a wide range of power—typically 24 VDC, 120 VAC, and 240 VAC. Some manufacturers produce units that will operate on any voltage from 120 VAC to 240 VAC without any modifications to the unit. Connection of power to the dc-type units requires that careful attention be paid to ensure that the positive (+) and negative (−) power wires are correctly connected. Failure to do so can result in serious damage to the PLC. Power connection to ac units is not as critical unless the PLC specifications may require specific connection of the hot and neutral wires to the proper terminals. However, no matter which style PLC is being utilized, proper fuses must be inserted in the power line connections to protect both the PLC and the power wiring from overcurrent either from accidental shorts or equipment failure. The installation manual for the particular PLC being used will generally provide fusing information for that unit. The wiring diagram for an ac-type PLC is shown in Figure 6–1.

Notice that incoming power is first connected to a disconnect switch. When turned off, this switch will disconnect all power from the fuses and the PLC. This provides safety for personnel performing maintenance on the system by totally removing power from the system. The neutral conductor is typically grounded at the source. If the neutral is grounded, it does not need to be included in the disconnect switch. However, if the neutral ground is lost, it would be possible to receive a shock if the neutral wire was touched. For safety, it is always better to totally disconnect power. Two fuses are shown, one for hot and one for neutral. Again, if the neutral is grounded, the neutral fuse is not required. However, for the same reason as the disconnect, the second fuse is desirable since it will protect the system against heavy neutral current that could result if the ground is lost.

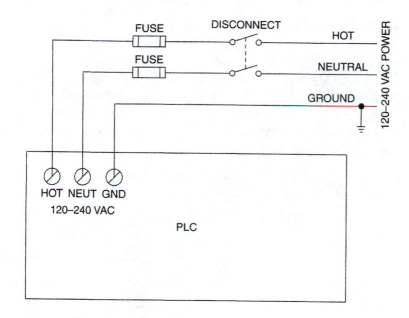

FIGURE 6–1

Typical ac Power Wiring

Some discussion of the terms **hot** and **neutral** is required here. Utility power is generated as a three-phase (3φ) voltage. This is accomplished by the wiring scheme in the generator that produces three voltage sources at a phase angle of 120° from each other. The schematic representation of this type of generator is shown in Figure 6–2. Notice that the generator has three windings with the outputs labeled PHASE A, PHASE B, and PHASE C. There is also a fourth terminal on the generator labeled NEUTRAL, which is connected to the common connection of all three phase terminals. For this discussion, assume we are using 120/208 VAC 3φ. If the generator of Figure 6–2 were producing this voltage, the voltage from any PHASE (A, B, or C) to NEUTRAL would be 120 VAC, and the voltage between any two phases (PHASE A/PHASE B, PHASE B/PHASE C,

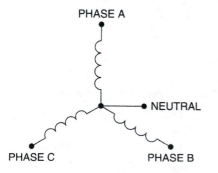

FIGURE 6–2

Three-Phase Generator Output Schematic

or PHASE C/PHASE A) would be 208 VAC. The three phase leads are referred to as the HOT leads, and the common connection to all three phase windings is referred to as the NEUTRAL lead. In practice, the NEUTRAL connection is connected to earth ground at the generator. This is true in residential and commercial buildings with 120 VAC power. The NEUTRAL wire in the building is connected to earth ground at the panel where power enters the building. For this reason, if a voltmeter were placed between the NEUTRAL wire and the safety ground wire in any receptacle, the voltage read would be close to or at 0 VAC.

In some cases, PLCs are operated from dc power instead of ac power. Figure 6–3 illustrates the power connection for a PLC requiring dc power, in this case, 24 VDC.

FIGURE 6–3

Typical dc Power Wiring

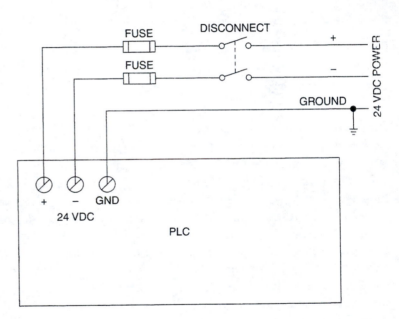

This wiring also includes fusing and disconnecting for both power conductors. If the (−) power line is grounded at the source, the (−) disconnect and fuse would not be required. However, as with the ac power wiring, it is always safer to provide for fusing and disconnection of both power conductors.

Care must be taken to ensure that the wiring is properly connected to avoid damage to the equipment and to the personnel coming in contact with it. For this reason, in this chapter, a very simplistic approach will be taken to describe wiring techniques so that it will explain the wiring requirements thoroughly enough to allow all readers to understand the principles associated with properly connecting the PLC to the system.

To connect power to the PLC, the PLC may be thought of as a lightbulb that needs to be lit; the two power wires are connected to the two wires of the light-bulb and must be insulated from each other. In the case of a PLC operating on dc power, it may be thought of as an LED. For a lightbulb, no matter which power wire connects to which lightbulb wire, the light will still light. This is also true for an ac-powered PLC. It generally does not matter which power wire is connected to the power terminals as long as both are connected and insulated from each other. In the case of the LED, though, the positive (+) and negative (−) connections must be made to the proper LED wires and insulated from each other if the LED is to light. The same is true for the dc-powered PLC. The difference with the PLC is that if it is connected incorrectly, the damage can be very expensive.

6-2 Input Wiring

The inputs of modern PLCs are generally opto-isolators. An opto-isolator is a device consisting of a light-producing element, such as an LED, and a light sensing element, such as a phototransistor. When a voltage is applied to the LED, light is produced which strikes the photo-detector. The photo-detector then changes its output (e.g., in the case of a phototransistor, it saturates). The separation of the sensing and output devices in the opto-isolator provides the input to the PLC with a high voltage isolation since the only connection between the input terminal and the input to the PLC is a light beam. The light-producing element and any current-limiting device and protection components determine the input voltage for the opto-isolator. For instance, an LED with a series current-limiting resistor would be sized to accept 5 VDC, 24 VDC, or 120 VDC. To accept an ac signal, two opposing LEDs with a series current-limiting resistor are used. The resistor would be sized to allow the LED to light with 5 VAC, 24 VAC, 120 VAC, 240 VAC, or any voltage we desire. PLC manufacturers offer different models having various input voltage specifications. The PLC with the input voltage specification is chosen at the time of purchase.

Figure 6-4 shows two types of opto-isolators utilized. The dc unit is shown in (a) and the ac unit in (b). The wires from the switch or sensor are connected to the left side of the drawing. The right side of the device is connected internally to the actual PLC input. Notice that each opto-isolator has a series resistor to limit the device current. Also notice that the ac unit has opposing LEDs inside the device so light is produced on both half cycles of the input voltage.

Since the input to the PLC is an LED, we can visualize the wiring of the input by thinking of it as some type of device controlling a lightbulb and requiring that

FIGURE 6–4

*Typical PLC Input
Circuits*

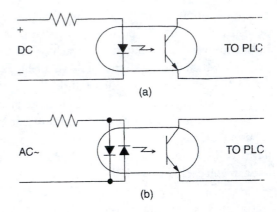

the input device lights the lightbulb. The input device may be a switch or some type of on/off sensor such as a photosensor or proximity sensor, but the goal is always the same: Wire the device so the LED will light when we want the input to be detected as ON. For the dc unit, note that the proper polarity must be observed for the LED to light. In the case of the ac unit, since there is a forward-biased LED, no matter which way current flows, polarity does not matter. In fact, some manufacturers produce PLCs with ac/dc inputs that are actually ac units. When using this type of PLC, if the input signal is dc, the polarity does not matter because one of the LEDs will light no matter which voltage polarity is applied.

PLC inputs are configured in one of two ways. In some units, all inputs are isolated from each other; that is, there is no common connection between any two inputs. Other units have one side of each input connected to one common terminal. The PLC brand and model determines whether the power for all input devices is common. The power supply for the inputs may be either external or internal to the PLC. In low voltage units (24 VDC), a power supply capable of supplying enough current to turn on all inputs will be internal to the PLC. If all inputs will be from switches, no other power supply will be required for the inputs. This internal power supply may not be large enough to also supply any active sensors (photodetectors or proximity detectors) that may be connected. If not, an external supply will need to be obtained. The PLC's specifications will indicate the capacity of this internal power supply. The internal schematic for the inputs of a PLC having three inputs with common connection is shown in Figure 6–5. Notice that all three opto-isolators have one wire connected to the same terminal, the INPUT COM. Also, the wire that is connected to the INPUT COM terminal for each opto-isolator is the negative connection for lighting the LEDs in the opto-isolators. This means that any switch device connected to terminals INPUT1

INPUT2, and INPUT3 must have the opposite end of the device connected to a positive voltage in order to light the LED in the opto-isolator. If more than one power supply is used to power the devices, all of them must have the negative power lead connected to INPUT COM because that is the only terminal available to connect to the negative input of the opto-isolators.

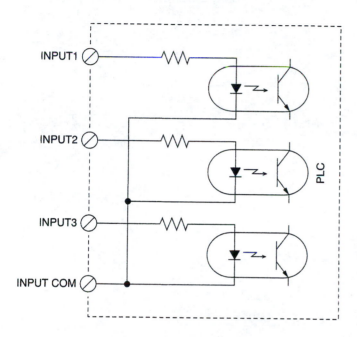

FIGURE 6–5

PLC with Common Inputs

A PLC may also have isolated inputs, as shown in Figure 6–6. Since each input has no connection with any other input, each one may be connected as desired with no concern for power supply interaction. The only requirement is that, for the LED in the opto-isolator to light, a positive voltage must be applied to the positive (+) terminal of the PLC input with respect to the (−) terminal.

6-3 Inputs Having a Single Common

Let us now examine the wiring connections required to implement a system in which all inputs have a common power supply. For this type of system, a PLC with inputs having a single common connection may be used, as shown in Figure 6–5. A system of this type is shown in Figure 6–7.

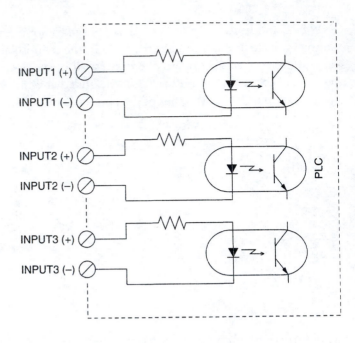

FIGURE 6–6

PLC with Isolated Inputs

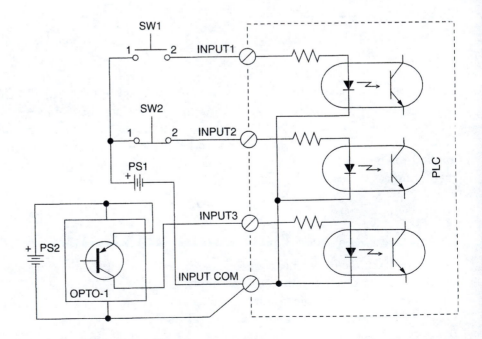

FIGURE 6–7

Non-Isolated Input Wiring

This drawing shows three devices connected to the PLC, a normally open pushbutton (SW1), a normally closed pushbutton (SW2), and a photodetector (OPTO-1). The system also has two power supplies, PS1 and PS2, providing power for the input devices. Notice that the two power supplies have the negative side connected to the INPUT COM terminal. Let us first examine the connection for normally open pushbutton SW1. Remember to think of the input in terms of lighting the LED in the opto-isolator. Since the LED must be forward-biased to light, the positive voltage must be applied to the anode of the LED and the negative voltage to the cathode. All three opto-isolators have the cathode lead of the LED connected to the INPUT COM terminal. Therefore, any power supply used to light the LEDs must have the negative lead connected to the INPUT COM terminal. For this reason, both PS1 and PS2 have the negative lead of the supply connected to the INPUT COM terminal. With the negative lead of PS1 connected to the cathode of the INPUT1 LED, the positive lead of PS1 must be connected to the anode of the INPUT1 LED. If this is done, the INPUT1 LED will light, and the PLC will accept INPUT1 as being turned ON. To control this INPUT, SW1 is placed in series with the positive power supply lead going to INPUT1. If SW1 is not pressed, the switch will be open (remember it is a normally open switch) and no voltage will be applied across the INPUT1 LED. The LED will not light, and the PLC will accept INPUT1 as being OFF. If SW1 is pressed, the SW1 contacts will close, and the positive lead of PS1 will be connected to the anode of the INPUT1 LED causing the LED to light. This will cause the PLC to accept INPUT1 as being ON.

INPUT2 is connected similar to INPUT1 except that switch SW2 is a normally closed pushbutton. Since it is normally closed if the switch is not being pressed, the positive lead of PS1 will be connected to the INPUT2 LED, causing it to light. This will cause the PLC to accept INPUT2 as being ON. When pushbutton switch SW2 is pressed, the normally closed contacts of the switch will open and remove power from the INPUT 2 LED. With the LED not lit, the PLC will accept INPUT2 as being OFF.

INPUT3 is connected to photodetector OPTO-1. OPTO-1 is a photodetector with a PNP transistor output. In operation, OPTO-1 requires dc power, and for that reason PS2 is connected to the device. Notice that the emitter of OPTO-1 is connected to the positive terminal of PS2 and that the collector is connected to INPUT3. When light strikes the phototransistor in OPTO-1, the transistor saturates and the resistance from collector to emitter becomes very low. When the transistor saturates, INPUT3 is pulled to the positive terminal voltage of PS2. This will cause the LED for INPUT3 to light since it will have positive voltage on the anode and negative on the cathode. When the LED lights, the PLC will accept INPUT3 as being ON. When no light strikes the phototransistor in OPTO-1, the transistor will switch off, presenting a high resistance from collector to emitter. This high resistance will prevent current from flowing in the INPUT3 LED, and the LED will not light. With the LED not lit, the PLC will accept INPUT3 as being OFF.

A potential problem exists at INPUT2. SW2 is a normally closed pushbutton, and power is always applied to INPUT2. If power supply PS1 should fail, the PLC would assume that SW2 had been pressed since the LED for INPUT2 would not be lit. For this reason, it is good practice to use normally open switches and other devices to prevent a false decision by the PLC on the condition of the input device. One exception to this general rule occurs in the case of "fail safe" design, which will be discussed in a later chapter.

6-4 Isolated Inputs

Figure 6–8 illustrates a system utilizing a PLC with isolated inputs. This system has three power supplies, one for each input device. INPUT1 is supplied by power supply PS1 with normally open pushbutton switch SW1. The negative terminal of power supply PS1 is connected to the cathode terminal of the opto-

FIGURE 6–8

Isolated Input Wiring

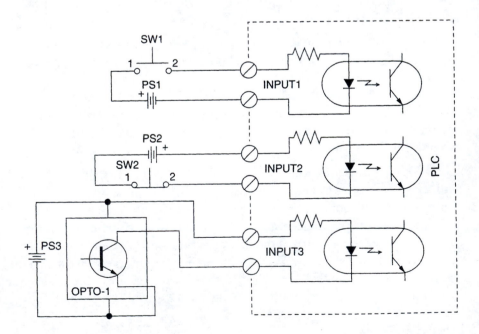

isolator for INPUT1. The positive terminal of PS1 is connected to the anode of the LED for INPUT1 through SW1. When SW1 is pressed, the current flow lights the LED, and the PLC accepts this as INPUT1 ON. With the switch SW1 released, the normally open contacts are open, no power is applied to the LED, and the PLC accepts INPUT1 as being OFF.

INPUT2 is connected to normally closed pushbutton switch SW2 and power supply PS2. In this case, however, the positive terminal of PS2 is permanently connected to the anode terminal of the LED for INPUT2. The cathode lead of the LED is connected to the negative terminal of PS2 through SW2. Since SW2 is normally closed, power will normally be applied to the LED for INPUT2, and the LED will be lit. This will cause the PLC to accept INPUT2 as being ON. When normally closed pushbutton switch SW2 is pressed, power is removed from the LED and it does not light. This causes the PLC to accept INPUT2 as being OFF. The potential problem noted in the previous paragraph associated with the use of a normally closed switch also exists here.

INPUT3 is connected to a photodetector OPTO-1 with an NPN transistor output. The positive terminal of power supply PS3 is connected directly to the anode of the LED for INPUT3. The collector of the phototransistor is connected to the cathode terminal of the LED, and the emitter of the phototransistor is connected to the negative terminal of power supply PS3. With this configuration, when the phototransistor saturates, the cathode of the LED for INPUT3 is pulled to the negative terminal of power supply PS3 causing current to flow in the LED. With the LED lit, the PLC will accept INPUT3 as being ON. When the phototransistor turns off, current through the LED stops flowing, the LED does not light, and the PLC accepts INPUT3 as being OFF.

Notice that with isolated inputs, wiring can be arranged in different ways to suit different requirements. Generally, if the PLC has an available internal power supply that has the current capacity to power the external input devices, external power is not required unless: (1) one of the input devices requires a power supply voltage that is different from the PLC input or (2) if the total current required by the device is more than the PLC can deliver.

Figure 6–9 illustrates the wiring diagram for a system with two normally open pushbutton switches and one photoelectric sensor connected to a PLC with 24 VDC inputs and an internal 24 VDC power supply. The power supply in this case is able to supply enough current to operate all three inputs and power the photoelectric sensor. Notice that the negative output of the internal power supply is connected directly to the INPUT COM of the input unit. The positive terminal of the internal power supply is connected to the two pushbutton switches and the power pin and emitter of the photoelectric sensor. INPUT3 is connected to the collector of the photoelectric sensor to allow the sensor to pull INPUT3 up when active. This also prevents any problem with loss of power since the emitter of the sensor would be open on power loss resulting in the input being OFF. For most applications, when possible, all inputs should be connected to input devices in such a manner as to cause the inputs to be normally OFF, except as previously noted.

FIGURE 6–9

Typical PLC Wiring Diagram

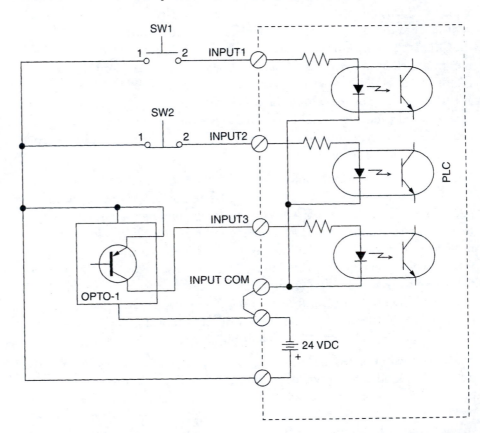

6-5 Output Wiring

PLC outputs are of two general types—relay or solid state. Relay outputs are magnetically operated mechanical switch contacts. Solid-state outputs may take the form of transistor or TTL logic (dc) and triac (ac). Relay outputs are usually used to control up to 2 amperes or when a very low resistance is required. Transistor outputs are open collector (common emitter), and use either NPN or PNP transistors. This type of output can control lamps and low power dc circuitry such as small dc relays. TTL logic outputs are available to drive logic circuitry. Triac outputs are used to control low power ac loads such as lighting, motor starters, and contactors. As with input units, output units are available either with a common terminal or isolated from each other. The type of output unit selected will determine the number of outputs being controlled and the power available for controlling those devices. Typically, power for driving output devices must be separately provided since there can be a wide range of requirements depending upon the device.

6-6 Relay Outputs

As stated before, relay outputs are normally used to control moderate loads (up to about 2 amperes) or when a very low on resistance is required. Refer to Chapter 1 for a description of a relay and a discussion of its operation. Relay contacts are described as three main arrangements or **forms,** which are FORM A, FORM B, and FORM C. A FORM A relay contact is a single pole normally open contact. This is analogous to a single contact normally open switch. The FORM B relay contact is a single pole normally closed contact that is similar to a single normally closed switch. The FORM C relay contact is a single pole double throw contact. The schematic symbols for the three arrangement types are shown in Figure 6–10.

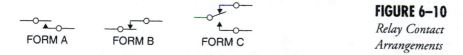

FORM A   FORM B   FORM C

FIGURE 6–10
Relay Contact
Arrangements

PLC output units are available with all three contact arrangements, but FORM A and FORM C are typically used. By specifying a FORM C contact, both FORM A and FORM B can be obtained by using either the normally open portion of the FORM C contact as a FORM A contact or by using the normally closed portion of the FORM C contact as a FORM B contact. Relay outputs are also available either with a common terminal or isolated contacts. An output unit with three FORM C contacts having a common terminal is shown in Figure 6–11. Note in this figure that the common terminal of each of the three relays is connected to one common terminal of the output unit labeled OUTPUT COM. Since all relays have one common terminal, all power supplies (there can be one or several) associated with the outputs to be driven must have one common connection. Also note that each relay has two labeled outputs, N/C (normally closed) and N/O (normally open).[1] The N/C and N/O labels have a number that indicates the

[1] Generally, the labels N/O and N/C are written with a slash; however, to save space, manufacturers usually omit the slash when marking the terminals on their products. The reader should consider the two labeling methods interchangable.

number of the output relay associated with the terminal. When an output is turned OFF, the OUTPUT COM terminal is connected to the N/C terminal associated with that output. When the output is turned ON, the OUTPUT COM terminal is connected to the associated N/O terminal.

FIGURE 6–11

Form C Relays with Common Output

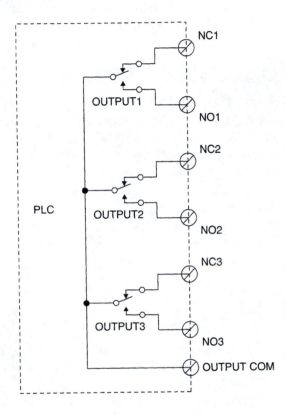

A typical connection diagram for a relay output unit with three FORM C contacts having common output is shown in Figure 6–12. In this drawing, the power source shown is an ac supply that could be 120 VAC power. Notice that the wiring of this figure shows lamp LT1 as only lighting when OUTPUT1 is turned ON. This is because the lamp is connected to the N/O terminal for OUTPUT1. Lamp LT2, however, is connected to the N/C terminal for OUTPUT3. This lamp will be ON whenever OUTPUT3 is turned OFF. This means that if the PLC were to lose power, lamp LT2 would light since there would be no power to energize the output relays in the PLC. In the ladder diagram, OUTPUT3 could be programmed as an always ON coil. The result would be that while the PLC was powered and running, lamp LT2 would not be lit. If the PLC lost power, lamp LT2 would light and provide the operator with an indication that the PLC had a problem. This method could also be provided as a tool to allow maintenance to troubleshoot and repair the system faster. Also in the drawing of Figure 6–12, coil K1 is shown connected to OUTPUT2. This could be a solenoid or the coil of a motor starter used to control power to a motor that requires more current than the relay in the output unit can safely carry. Note that to be used in this situation, the coil K1 would have to be rated for ac use at the voltage available from the ac power source. The wiring of these outputs may each be thought of in terms of a switch controlling a lamp. A normally closed switch or a normally open switch may be used. The switch is placed in series with the lamp and the power source to control current to the lamp.

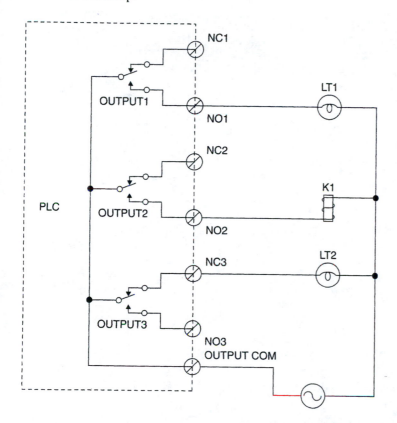

FIGURE 6–12

Common Relay Wiring

A PLC relay output unit with three isolated FORM C contacts is shown in Figure 6–13. In this type of output unit, the relay contacts have no connection between them. These output contacts may be used for any purpose to drive any devices with no concern for connection between power sources. Each output has three terminals. The C terminal is the common terminal of the relay. The N/O terminal is the normally open contact and the N/C terminal is the normally closed contact of the relay. The N/C and N/O labels have a number that is the associated output number. The same number is indicated for each C terminal as well as indicating that it is associated with a particular output. As with the common relay output unit of Figure 6–11, the N/O contact only *closes* when the output is ON and the N/C contact only *opens* when the output is ON.

FIGURE 6–13

Isolated Relay Output

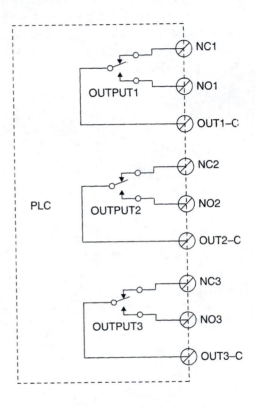

Figure 6–14 shows a typical system output wiring diagram using an output unit with three FORM C isolated outputs. In Figure 6–14, the three outputs are controlling devices with three different power requirements. OUTPUT1 is controlling a dc lamp, LT1, which has its own dc power source. Notice that lamp LT1 will be lit whenever OUTPUT1 is turned OFF. This could be a fault indicator for the system. Lamp LT2 is an ac lamp having its own ac source. LT2 is also lit when OUTPUT3 is turned OFF. OUTPUT2 is connected to a release valve that has an internal power source and only needs a contact closure to release. In this case, the release valve is connected to the normally closed terminal for OUTPUT2. This connection provides for the release valve to be in the release condition should the PLC lose power. This may be the requirement to provide for the machine being in a safe condition in case of system failure. Notice that in the wiring

of OUTPUT1 and OUTPUT3, the wiring provides for a power source, a relay, and a light, all in series, with the relay controlling the flow of current to the light.

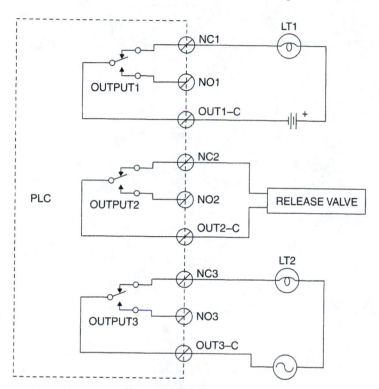

FIGURE 6–14

Isolated Contact Wiring

6-7 Solid State Outputs

There are several types of solid state outputs available with PLCs. Three popular types are transistor, triac, and TTL. All three of these output units will generally have a common terminal although triac output units are available in an isolated configuration. Transistor output units are usually open collector with the common terminal connected to the emitters of all outputs. A transistor output unit providing three open collector outputs is shown in Figure 6–15. In most units, the transistors are optically isolated from the PLC. This unit has all the emitters of the output transistors connected to one common terminal labeled OUTCOM. There are two different types of transistor units available. The NPN units, as shown in Figure 6–15, are referred to as **sinking,** and the PNP units, as shown in Figure 6–16, are referred to as **sourcing.** For sinking (NPN) outputs, current will always flow into the output terminal. For sourcing (PNP) outputs, current will always flow out of the output terminal.

FIGURE 6–15
*Transistor Sinking
Output*

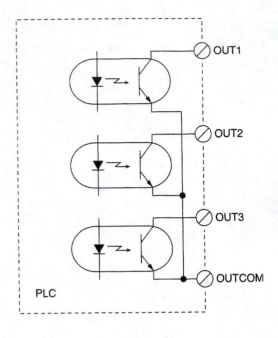

FIGURE 6–16
*Transistor Sourcing
Output*

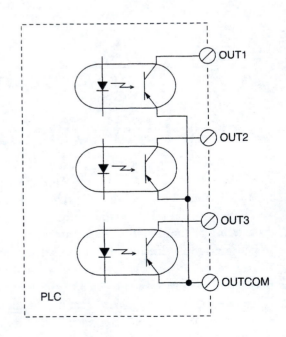

Another way of remembering the difference is that a sinking output will pull the output voltage in a negative direction, and the sourcing output will pull the output voltage in a positive direction. The sourcing and sinking description conforms to conventional current flow from positive to negative. Notice that the transistors used in Figure 6–16 are PNP type and that the common terminal would need to be connected to the positive terminal of the power supply in order for the transistor to be properly biased. This will cause the outputs to be pulled in a positive direction when the transistor is turned ON (a sourcing output).

A wiring diagram for a transistor sinking output unit is shown in Figure 6–17. This diagram shows three output devices connected to the output unit. Lamp LT1 will light when OUTPUT1 turns ON since the output transistor will saturate when the output turns ON. The saturated transistor will *sink* current to the OUTCOM terminal causing current to flow in the lamp. OUT2 is connected to coil K1. This could be a solenoid or relay. This device will be energized when OUTPUT2 is turned ON causing the output transistor to saturate. Notice that a diode, CR1, is connected across K1 in such a manner as to be normally reverse biased. This diode prevents excessive voltage buildup across K1 when the output

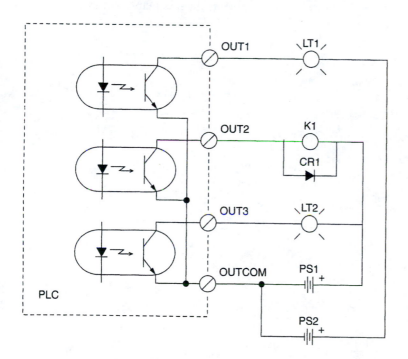

FIGURE 6–17

Transistor Output Wiring

transistor turns OFF. When the output transistor turns OFF and the magnetic field in the coil begins to collapse, a voltage in opposition to the applied voltage is developed. Unchecked, this voltage could be as high as several hundred volts. A voltage this high would quickly destroy the output transistor. The voltage is not allowed to build up because the current developed by the collapsing field is shunted through the diode. Transistor output units now generally have this diode built into the output unit for protection. Also, some relays are manufactured with this diode installed.

Notice that the diagram of Figure 6–17 includes two separate power sources, one providing power for LT1 and the other providing power for K1 and LT2. This is a typical situation since there may be occasions when devices connected to the output unit operate at different voltages. In this case, LT1 could be a 5-volt lamp, and LT2 and K1 could be 24-volt devices. The only requirement is that the two power supplies (PS1 and PS2) must have a common negative terminal that is connected to the OUTCOM terminal, and the voltage of PS1 and PS2 do not exceed the maximum rated voltage of the PLC output transistors.

Figure 6–18 contains the schematic drawing for a triac output unit. All output triacs have one common terminal that will be connected to one side of the ac power source providing power for the output devices being controlled. Each

FIGURE 6–18
Triac Output Unit

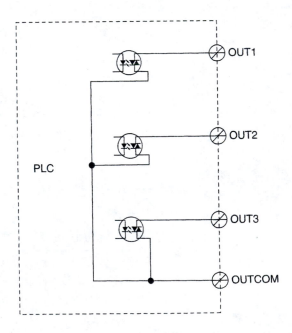

triac is triggered when the output associated with it is turned ON. There are units available that have zero crossover networks built in to provide for noise reduction by only allowing the triac to turn ON at the time the power signal crosses through zero volts. Noise and current spikes can be generated when the triac turns ON if the ac voltage is at some voltage other than zero. This is because the voltage to the output device being controlled will instantly go from zero with the triac OFF to the instantaneous value of the ac voltage at turn-on. If the triac turns ON at the time the ac voltage is passing through zero volts, no such spikes will be produced. Notice that the triac outputs shown in Figure 6–18 are optically isolated from the PLC. This allows the PLC to control very high voltage levels (120–240 VAC) with isolation of these voltages from the low voltage circuitry of the PLC.

Figure 6–19 contains the wiring diagram for a triac output unit. As can be seen in the diagram, the common terminal for the triacs is connected to one side of the ac source powering the output devices controlled by the unit. Output units are available with a variety of voltage and current ratings. The devices being controlled by the output unit shown in Figure 6–19 are the same as those in Figure 6–17; however, the difference is that all devices in Figure 6–17 are dc powered and all devices in Figure 6–19 operate on ac. Also notice that the diode across K1 is not present in the triac output unit drawing. The reason is that because we are using ac in this case, the diode would appear as a short circuit on alternating half cycles of the

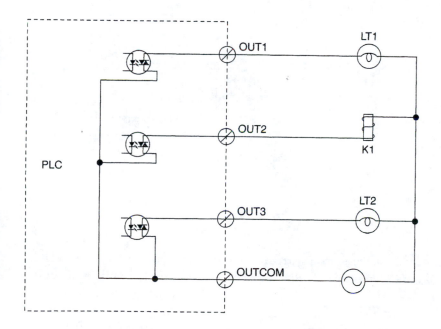

FIGURE 6–19

Triac Output Wiring

ac waveform. Since the triac turns OFF when the circuit current crosses through zero, there will be little or no magnetic field present in the relay coil.

Another popular output type is the TTL output. Figure 6–20 shows the schematic diagram of an output unit containing three TTL outputs. Notice that these outputs are also normally optically isolated. The outputs of the TTL unit have a common terminal (OUTCOM), which must be connected to the negative terminal of the power supply for the external TTL devices being driven by the output unit. Some TTL output units require that the 5 VDC power for the output circuitry be connected to the output unit to provide power for its internal TTL circuitry. The connection of these outputs to external TTL circuitry would be the same as with any other TTL connections. However, the designer needs to be aware that the logical output polarity (positive or negative logic) varies with different PLC manufacturers. How the output logical polarity is handled in the ladder diagram will be affected by this specification.

FIGURE 6–20

TTL Output Unit

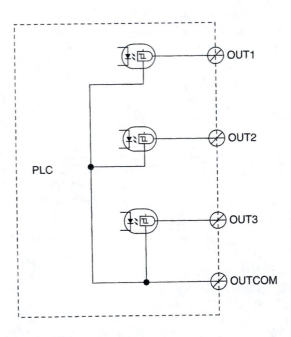

As can be seen from the above discussion of input and output units, there are a large variety of options available. Wiring for each type of unit is critical to the unit's proper operation. Much care must be taken to ensure proper operation of the input or output unit without causing damage. Also, there are safety concerns which must be addressed when planning and implementing the wiring of the system to provide for a system that will not be a hazard to the people operating and maintaining it.

Summary

As we have seen illustrated in this chapter, the proper application of a PLC into a control system involves much more than the writing of the software. Although being able to devise clever and elegant ladder logic programs is an important aspect of designing a PLC control system, the designer may often find that selecting a PLC, determining which types of input and output capabilities to purchase, and planning how the PLC should be wired into the system can consume as much of the design effort as the software itself. Even though PLC manufacturers have attempted to make this selection and application process as simple as possible, knowledge of the various types of available equipment is a valuable tool to best assure the success of the project.

Review Questions

1. When wiring power to a PLC, why are the fuses located on the switched side of the line?

2. What is the main difference between a dc opto-isolated input and an ac/dc opto-isolated input on a PLC?

3. What generally determines the voltage levels required by an opto-isolated input?

4. What is the limitation in using a PLC's internal power supply to provide power for the PLC inputs?

5. Cite one advantage in using isolated inputs as opposed to inputs sharing an "INPUT COM" terminal.

6. Draw the input wiring diagram for a PLC system using a 24 VDC input unit with the following inputs:

   IN1—Normally open pushbutton ON switch

   IN2—Normally open pushbutton OFF switch

   IN3—Normally closed selector switch labeled STEP 1/STEP 2

   IN4—Normally open mushroom-head palm switch

   Show all switches using the correct drawing symbol, and show all devices including the power supply. The PLC being used does not have an internal 24 VDC power supply.

7. Explain the difference between FORM A, FORM B, and FORM C relay contact arrangements.

8. Generally speaking, PLC outputs are not used to directly control the power provided to dc or ac motors. Instead, one PLC output is used to control a contactor (relay) that, in turn, switches power to the motor. Why is this?

9. Cite one advantage that relay outputs have over solid state outputs.

10. Of the two types of transistor output configurations (PNP and NPN), which is sourcing and which is sinking?

11. If one terminal of a lamp is connected directly to the negative terminal of the power supply, what kind of transistor output unit will be required (sourcing or sinking) to allow the PLC to light the lamp?

12. The negative lead of a power supply is connected directly to one lead of a coil. Draw the wiring diagram for the proper connection of the coil to a transistor output unit. Also, show the spike protection diode across the coil.

13. Cite one of the advantages that triac outputs have over relays when switching ac loads.

14. Using a relay output unit with one FORM C contact driven from OUTPUT1, draw the wiring diagram for a system with two ac powered lamps, L1 and L2. L1 will light when OUTPUT1 is ON, and L2 will light when OUTPUT1 is OFF.

Analog I/O

Objectives

Upon completion of this chapter, you will know:

- the operations performed by analog-to-digital (A/D) and digital-to-analog (D/A) converters.

- some of the terminology used to describe analog converter performance parameters.

- how to select a converter for an application.

- the difference between a unipolar and bipolar converter.

- some common problems encountered with analog converter applications.

Introduction

Although most of the operations performed by a PLC are either discrete I/O or register I/O operations, there are some situations that require the PLC to either monitor an analog voltage or produce an analog voltage. As an example of an analog monitoring function, consider a PLC that is monitoring the wind speed in a wind tunnel. In this situation, an air flow sensor is used that outputs a dc voltage that is proportional to wind speed. This voltage is then connected to an analog input (also called A/D input) on the PLC. As an example of an analog control function, consider an ac variable frequency motor drive (called a VFD). This is an electronic unit that produces an ac voltage with a variable frequency. When connected to a three-phase ac induction motor, it can operate the motor at speeds other than rated speed. VFDs are generally controlled by a 0–10 volt dc analog input with zero volts corresponding to zero speed and 10 volts corresponding to rated speed. PLCs can be used to operate a VFD by connecting the analog output (also called D/A output) of the PLC to the control input of the VFD.

Analog input and output values are generally handled internally by the PLC as register operations. Input and output transfers are usually done at update time unless otherwise specified by the programmer. Internally, the values can be manipulated mathematically and logically under ladder program control.

7-1 Analog (A/D) Input

Analog inputs to PLCs are generally accomplished using add-on modules, which are extra cost items. Few PLCs have analog input as a standard feature. Analog inputs are available in unipolar (positive input voltage capability) or bipolar (plus and minus input voltage capability). Typical, standard, off-the-shelf unipolar analog input modules have ranges of 0–5 VDC and 0–10 VDC, while standard bipolar units have ranges of −5 to +5 VDC and −10 to +10 VDC.

Specifying an Analog Input

There are basically three characteristics that need to be considered when selecting an analog input. They are described in the following paragraphs.

Unipolar (positive only) or Bipolar (plus and minus)   This is a simple decision. If the voltage being measured will never be negative, then a unipolar input is the best choice. It is not economical to purchase a bipolar input to measure a unipolar signal.

Input Range   This is relatively simple also. For this specification, you will need to know the type of output from the sensor, system, or transducer being measured. If you expect a signal greater then 10 volts, purchase a 10 volt input and divide the voltage to be measured using a simple resistive voltage divider (keep in mind that, if necessary, you can restore the value in software by a simple multiplication operation). If you know the measured voltage will never exceed 5 volts, avoid purchasing a 10 volt converter because you will be paying extra for the additional, unused range.

Number of Bits of Resolution   The resolution of an A/D operation determines the number of digital values the converter is capable of discerning over its range. As an example, consider a unipolar analog input with 4 bits of resolution and a 0–10 volt range. With 4 bits, we will have 16 voltage steps, including zero. Therefore, zero volts will convert to binary 0000 and the converter will divide the 10 volt range into 16 increments. It is important to understand that with four binary bits, the largest number that can be provided is 1111_2 or 15_{10}. Therefore, the largest voltage that can be represented by a 10 volt, 4-bit converter is $10 \times 15/16 = 9.375$ volts. In other words, our 10 volt converter is incapable of measuring 10 volts. All converters are capable of measuring a maximum voltage that is equal to the rated voltage (sometimes called V_{REF}) times $(2^n - 1)/(2^n)$, where n is the number of bits. Since our converter divides the 10 volt range into 16 equal parts, each step will be $10/16 = 0.625$ volt. This means that a binary value of 0001 (the smallest increment) will correspond to 0.625 volt. This is called the **voltage resolution** of the converter. Sometimes we refer to resolution as the number of bits the converter outputs, which is called the **bit resolution.** Our example converter has a bit resolution of 4 bits. The bit resolution (and voltage resolution) of an A/D converter determines the smallest voltage increment the converter can determine. Therefore, it is important to be able to properly specify the converter. If we use a converter with too few bits of resolution, we will not be able to correctly measure the input value to the degree of precision needed. Conversely, if we specify too many bits of resolution, we will be spending extra money for unnecessary resolution.

For a unipolar converter, the voltage resolution is the full scale voltage divided by 2^n, *where n is the bit resolution.* As a rule of thumb, you should select a converter with a voltage resolution that is approximately 25 percent or less of the desired resolution. Increasing the bit resolution makes the voltage resolution smaller.

Consider Table 7–1 for a 4-bit, 10 volt unipolar converter. The leftmost column shows the sixteen discrete steps that the converter is capable of resolving, 0 through 15. The middle column shows the binary value. The rightmost column shows the corresponding voltage that is equal to the step times the rated voltage times $(2^n - 1)/(2^n)$. For our converter, this will be the step times 10 volts $\times$ 15/16. Again, notice that even though this is a 10 volt converter, the highest voltage that it can convert is 9.375 volts, which is one step below 10 volts.

Example 7–1
Problem:

A temperature sensor outputs 0–10 volts dc for a temperature span of 0–100° C. What is the bit resolution of a PLC analog input that will digitize a temperature variation of 0.1° C?

Solution:

Since, for the sensor, 10 volts corresponds to 100° C, the sensor outputs 10 V/100° C = 0.1 volt/° C. Therefore, a temperature variation of 0.1° would correspond to 0.01 volt, or 10 millivolts from the sensor. Using our rule of thumb, we would need an analog input with a voltage resolution of

10 mV × 25 percent = 2.5 mV (or less) and an input range of 0–10 volts. This means the converter will need to divide its 0–10 volt range into 10 V/2.5 mV = 4,000 steps. To find the bit resolution, we find the smallest value of n that solves the inequality $2^n > 4,000$. The smallest value of n that will satisfy this inequality is $n = 12$, where $2^{12} = 4,096$. Therefore, we would need a 12-bit, 10 volt analog input. Now, we can find the actual resolution by solving for a 12-bit, 10 volt converter. The resolution would be $10 \text{ V}/2^{12} = 2.44$ mV. This voltage step would correspond to a temperature variation of 0.0244 degree. This means that the digitized value will be within ± 0.0122 degree of the actual temperature.

Table 7-1 Output Table for 4-Bit, 10 Volt Unipolar A/D Converter

$Step_{10}$	$Step_2$	Voltage
0	0000	0.000
1	0001	0.625
2	0010	1.250
3	0011	1.875
4	0100	2.500
5	0101	3.125
6	0110	3.750
7	0111	4.375
8	1000	5.000
9	1001	5.625
10	1010	6.250
11	1011	6.875
12	1100	7.500
13	1101	8.125
14	1110	8.750
15	1111	9.375

Determining the number of bits of resolution for a bipolar converter utilizes a similar method. Bipolar converters generally utilize what is called an **offset binary** system. In this system, all binary zeros represent the largest negative voltage, and all binary ones represent the largest positive voltage minus one bit-resolution. To illustrate, assume we have an A/D converter with a range of −10 volts to +10 volts and a bit resolution of 8 bits. Since the overall range is 20 volts

(-10 to $+10$ volts), the voltage resolution will be 20 volts/2^8 = 78.125 mV. Therefore, the converter will equate 00000000_2 to -10 volts and 11111111_2 will become $+10$ V $-$ 0.078125 V = 9.951875 V. Keep in mind that this will make the binary number 10000000_2, or 128_{10} (called the half-range value), to be -10 V $+ 128 \times 78.125$ mV = 0.000 V.

Consider Table 7–2 for a 4-bit, 10 volt bipolar converter. The leftmost column shows the sixteen discrete steps that the converter is capable of resolving, 0 through 15. The middle column shows the binary value. The rightmost column shows the corresponding voltage, which is equal to the step times the voltage span times $(2^n - 1)/(2^n)$. For our converter, this will be the step times 20 volts $\times$ 15/16. Notice that digital zero corresponds to -10 volts, the half value point 1000_2 corresponds to zero volts, and the highest voltage that it can convert is 8.750 volts, which is one step below $+10$ volts.

Table 7–2 Output Table for 4-bit 10 Volt Bipolar A/D Converter

$Step_{10}$	$Step_2$	Voltage
0	0000	$-$ 10.000
1	0001	$-$ 8.750
2	0010	$-$ 7.500
3	0011	$-$ 6.250
4	0100	$-$ 5.000
5	0101	$-$ 3.750
6	0110	$-$ 2.500
7	0111	$-$ 1.250
8	1000	0.000
9	1001	1.250
10	1010	2.500
11	1011	3.750
12	1100	5.000
13	1101	6.250
14	1110	7.500
15	1111	8.750

It is important to understand that expanding the span of the converter (span is the voltage difference between the minimum and maximum voltage capability of the converter) to cover both positive and negative voltages increases the value of the voltage resolution, which in turn detracts from the precision of the converter. For example, an 8-bit, 10 volt *unipolar* converter has a voltage resolution of $10/2^8 = 39.0625$ mV, while an 8-bit, 10 volt *bipolar* converter has voltage resolution of $20/2^8 = 78.125$ mV.

Example 7–2
Problem:

A 10-bit bipolar analog input has an input range of -5 to $+5$ volts. If the converter outputs the binary number 0110111101_2, what is the voltage being read?

Solution:

First, we find the voltage resolution of the converter. Since the span is 10 volts (-5 to $+5$ volts), the resolution is $10/2^{10} = 9.7656$ mV. Next, we convert the output binary number to decimal ($0110111101_2 = 445_{10}$) and multiply it by the resolution to get $10/2^{10} \times 445 = 4.3457$ V. Finally, since the converter uses offset binary, we subtract 5 volts from the result to get 4.3457 V $- 5$ V $= -0.6543$ V.

The impedance of the source of the voltage to be measured must also be a consideration. However, this factor usually does not affect the selection of the analog input because generally all analog inputs are high impedance (1 Megohm or higher). Therefore, if the source impedance is also high, the designer should exercise caution to make sure the analog input does not load the source and create a voltage divider. This will cause an error in the reading. As a simple example, assume the voltage to be read has a source impedance of 1 Megohm and the analog input also has an impedance of 1 Megohm. This means that the analog input will only read half the voltage because the other half will be dropped across the source impedance. Ideally, a 1000:1 or higher ratio between the analog input impedance and the source impedance is desirable. Any lower ratio will cause a significant error in the measurement. Fortunately, since most analog sensors utilize operational amplifier outputs, the source impedance will generally be extremely low and loading error will not be a problem.

7-2 Analog (D/A) Output

When selecting an analog output for a PLC, the design considerations are almost the same as with the analog input. Most analog outputs are available in unipolar 0 to 5 V and 0 to 10 V, and in bipolar -5 to $+5$ V and -10 to $+10$ V systems. The methods for calculating bit resolution and voltage resolution are the same as for analog inputs so the selection processes are very similar.

However, one additional design consideration that must be investigated when applying an analog output is load impedance. Most D/A converters use operational amplifiers as their output amplifiers. Therefore, the maximum current capability of the converter is the same as the output current capability of the operational amplifier—typically about 25 mA. In most cases, a simple ohm's law calculation will indicate the lowest impedance value that the D/A converter is capable of accurately driving.

Example 7–3
Problem:

A 12-bit, 10 volt bipolar analog output has a maximum output current capability of 20 mA. It is connected to a load that has a resistance of 330 ohms. Will this system work correctly?

Solution:

If the converter were to output its highest magnitude of voltage, which is −10 volts, the current would be −10 V/330 ohms = −30.3 mA. Therefore, in this application, the converter will not work correctly because it would go into current limiting mode for any output voltage greater than 20 mA × 330 ohms = 6.6 V (of either polarity).

7-3 Analog Data Handling

As mentioned earlier, analog I/O is generally handled internally to the PLC as register values. For most PLCs, the values can be mathematically manipulated using software math operations. For more powerful PLCs, these can be done in binary, octal, decimal, or hexadecimal arithmetic. However, in the lower cost PLCs, the PLC may be limited in the variety of number systems it is capable of handling so the designer may be forced to work in a number system other than decimal. When this occurs, simple conversions between the PLC's number system and decimal will allow the designer to verify input values and program output values.

Example 7–4
Problem:

A voltage of 3.500 volts is applied to an 8-bit, 5 volt unipolar analog input of a PLC. Using monitor software, the PLC analog input register shows a value of 263_8. Is the analog input working correctly?

Solution:

First, convert 263_8 to decimal. It would be $2 \times 8^2 + 6 \times 8^1 + 3 \times 8^0 = 179_{10}$. For an 8-bit, 5 volt unipolar converter, 179 would correspond to $179 \times 5/2^8 = 3.496$ volts. Since the resolution is $5/2^8 = 0.0195$ volts, the result is within 1/2 of a bit and is therefore correct.

7-4 Analog Input Potential Problems

After installing an analog input system, it sometimes becomes apparent that there are problems. Generally, these problems occur in analog inputs and fall into three categories: a constant offset error, percentage offset error, or an unstable reading.

Constant Offset Error

In constant offset errors, the correct values always differ from the measured values by an additive (or subtractive) constant. It is also accompanied by a zero error (where zero volts is not measured as zero). Although there are many potential causes for this, the most common is that the analog input is sharing a ground circuit with some other device. The other device is drawing significant current through the ground such that a voltage drop appears on the ground conductor. Since the analog input is also using the ground, the voltage drop appears as an additive analog input. This problem can be avoided by making sure that all analog inputs are two-wire inputs and that both of the wires extend all the way to the source. Also, the negative (−) wire of the pair should only be grounded at one point (called **single point grounding**). Care should be taken here because many analog sensors have a negative (−) output wire that is grounded inside the sensor. This means that if you also ground the negative wire at the analog input, you will create the potential for a **ground loop** with its accompanying voltage drop and analog input error. In this case, you would allow the negative lead to be grounded only at the sensor (i.e., it would not be grounded at the analog input).

Percentage Offset Error

This type of error is also called gain error. This is apparent when the measured value can be corrected by multiplying it by a constant. It can be caused by a gain error in the analog input, a gain error in the sensor output, or most likely, loading effect caused by interaction between the output resistance of the sensor and the input resistance of the analog input. Also, if a resistive voltage divider is used on the input to reduce a high voltage to a voltage that is within the range of the analog input, an error in the ratio of the two resistors will produce this type of problem.

Unstable Reading

This is also called a noisy reading. It appears in cases where the source voltage is stable, but the measured value rambles, usually around the correct value. It is usually caused by external noise entering the system before it reaches the analog input. There are numerous possible reasons for this; however, they are all generally caused by electromagnetic or electrostatic pickup of noise by the wires connecting the signal source to the analog input. When designing a system with analog inputs (or troubleshooting a system with this type of problem), remember that the strength of an electromagnetic field around a current-carrying wire is directly proportional to the current being carried by the wire and the frequency of that current. If an analog signal wire is bundled with or near a wire carrying high alternating currents or high frequency signals, it is likely that the analog signal wires will pick up electrical noise. There are some standard design practices that will help reduce or minimize noise pickup.

1. For the analog signal wiring, use twisted pair shielded cable. The twisted pair will cause electromagnetic interference to appear equally in both wires, which will be cancelled by the differential amplifier at the analog input. The copper braid shield will supply some electromagnetic shielding and excellent electrostatic shielding. To prevent currents from circulating in the shield, ground the shield only on one end.

2. Use common sense when routing analog cables. Tying them into a bundle with ac line or controls wiring, or routing the analog wires near high current conductors or sources of high electromagnetic fields (such as motors or transformers) is likely to cause problems.

3. If all else fails, route the analog wires inside a steel conduit. The steel has a high magnetic permeability and will shunt most, if not all, interference from external magnetic fields around the wires inside, thereby shielding the wires.

Summary

The addition of an analog I/O to a PLC system adds greater capability and flexibility to the system. However, it is important to understand how the analog I/O operates, how to properly specify and select the I/O, and some of the potential problems that can be encountered (and how to avoid them). In this chapter, we investigated several variations of analog input and output systems; however, the reader should note that there are other less common types of systems that utilize different binary enumeration schemes than those discussed in this chapter. Although these systems are not difficult to understand, their variations will require the programmer to write the ladder code to work with these systems.

Review Questions

1. What is the voltage resolution of a 10-bit unipolar 5 volt analog input?

2. A 10 volt unipolar analog input has a voltage resolution of 78.125 mV. What is the bit resolution?

3. How many bits would be needed for an analog output if, after applying the 25 percent rule of thumb, we need a resolution of 4.8 millivolts for a signal that has a range of 0 to +10 volts?

4. A 10-bit, unipolar 10 volt analog input has +4.35 volts applied to the input terminals. What will be the converted value in decimal format? Hexadecimal format? Binary format? Octal format?

5. An 8-bit, bipolar 5 volt analog input has an input of −3.29 volts. What will be the decimal value of the number the converter sends to the CPU in the PLC?

6. A PLC has been set up to output the binary number 10110101 to its analog output. The analog output is 8-bits, 10 volts, and bipolar. What DC voltage would you expect to see on the output?

7. An ac motor controller has a frequency control input of 0–10 volts dc, which varies the output frequency from 0–60 Hz. It drives a three-phase induction motor that is rated at 1750 RPM at 60 Hz. The dc input to the motor controller is provided by an analog output from a PLC. The analog output is unipolar, 10 volts, 10 bits. What binary number must you program into the PLC for it to output the appropriate voltage to run the motor at 1000 RPM? (Assume for the motor that the relationship between frequency and speed is a simple ratio.)

8. A pressure sensor is rated at 0–500 psi and has an output range of 0–10 volts dc (10 volts corresponds to 500 psi). It is connected to a PLC's analog input that is 12 bits, 10 volts, and unipolar. If the PLC reads the analog input as $7B3_{16}$, what is the pressure in psi?

9. For the previous problem, if the analog value $7B3_{16}$ is converted to decimal and stored in an internal holding register, what constant value should you multiply it by to make the result appear as the actual reading in PSI? (Assume the PLC is capable of performing floating point math; that is, it can multiply by a decimal fraction.)

10. A liquid level sensor is installed in a 250 gallon tank such that it outputs zero volts when the tank is empty, 10 volts when the tank is full, and a proportional voltage from 0–10 volts for all levels in between. It will be connected to a PLC analog input that will allow the PLC to monitor the tank liquid level in increments of one gallon or less. Using the 25 percent rule-of-thumb, specify the least expensive analog input needed.

11. An 8-bit, 10 volt unipolar input on a PLC is designed to be connected to read the output of a 0 to 15 volt, 0 to 300 psi air pressure sensor. Upon inspection, we find that the person who designed the system connected a 50 percent voltage divider on the analog input of the PLC consisting of two precision 500 ohm resistors so that the 10 volt analog input could read the 15 volt sensor. If we include the 50 percent divider in the calculations, find the following: (a) the voltage resolution, (b) the full-scale voltage range, and (c) the constant multiplier that the PLC must use to convert the numerical output of the A/D converter into actual psi.

12. An 8-bit, 10 volt unipolar analog input on a PLC is not operating correctly. We connect a computer to the PLC and set up our monitoring software. Upon doing some measurements on the system, we find that if we input zero volts dc, the PLC indicates 00000101_2, and if we input 8 volts dc, the PLC indicates 11011100_2. What do you suspect is wrong with the system, and what suggestions would you make to someone troubleshooting it?

Discrete Position Sensors

Objectives

Upon completion of this chapter, you will know:

- the difference between a discrete sensor and an analog sensor.
- the theory of operation of inductive, capacitive, ultrasonic, and optical proximity sensors.
- which types of proximity sensors are best suited for particular applications.
- how to select and specify proximity sensors.

Introduction

Generally, when a PLC is designed into a machine control system, it is not simply put into an open loop system. This would be a system in which the PLC provides outputs but never "looks" to see if the machine is responding to those outputs. Instead, PLCs are generally put into a closed loop system. This is a system in which the PLC monitors the performance of the machine and provides the appropriate outputs at the correct times to make the machine operate properly, efficiently, and intelligently. In order to provide the PLC with a sense of what is happening within the machine, we use **sensors.**

In a way, a limit switch is a sensor. The switch senses when its actuator is being pressed and sends an electrical signal to the PLC input. The limit switch provides the PLC with a crude sense of touch. However, in many cases, the PLC needs to sense something more sophisticated than a switch actuation. For these applications, sensors are available that can sense nearly any parameter that may occur in a machine environment.

By no means does this text claim to contain a comprehensive coverage of all the currently available sensor technology. Because of the use of lasers, chroma recognition, image recognition, and other newer technologies, the state of sensor sophistication and variety of sensors is constantly evolving. Because of this rapid evolution, even experienced designers find it difficult to keep pace with sensor technology. Generally, machine controls and automation designers rely on manufacturers' sales representatives to keep them abreast of newly developing technologies. In many cases, a visit by a sales representative to view and discuss the potential sensor application will result in suggestions, catalogs, on-site demonstrations of sample units, and, if necessary, phone contact with a vendor's field engineer to further discuss the application. This network of sales representatives and applications engineers should not be ignored by the designer—they are a valuable resource. In fact, most of the material covered in this and subsequent chapters is provided by manufacturers' sales representatives.

8-1 Sensor Output Classification

Fundamentally, sensor outputs are classified into two categories—**discrete** (sometimes called digital, logic, or bang-bang) and **proportional** (sometimes called analog). Discrete sensors provide a single logical output (a zero or one). For example, a thermostat that operates the heating and air conditioning in a home is a discrete sensor. When the room temperature is below the thermostat's setpoint, it outputs a zero, and when the temperature rises above the setpoint, the thermostat switches on and provides a logical one output. It is important to remember that discrete sensors do not provide information about the current value of the parameter being sensed. They only decide if the parameter being sensed is above or below the setpoint. Again, using the thermostat example, if the thermostat is set for 70° F and it is off, all that can be concluded is that the room temperature is below 70° F. The room temperature could be 69° F, –69° F, or any other value below 70° F, and the thermostat will give the same logical zero output.

The schematic symbols for the discrete sensor are drawn as limit switches in a diamond-shaped box. In Figure 8–1, symbols (a) and (b) are respectively N/O and N/C sensors. Since this is a generic designation, we usually write a short description of the sensor type next to the symbol.

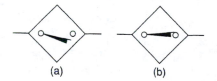

(a)   (b)

FIGURE 8–1

*Discrete Sensor
Schematic Symbol*

Proportional sensors, however, provide an analog output. The output may be a voltage, current, resistance, or even a digital word containing a discrete value. In any case, the sensor measures the value of the parameter, converts it to a signal that is proportional to the value, and outputs that value. When proportional sensors are used with PLCs, they are generally connected to analog inputs on the PLC instead of the discrete (digital) inputs. An example of a proportional sensor is the fluid level sending unit in the fuel tank of an automobile that sends a signal to operate the fuel level gage. This is generally a potentiometer in the fuel tank that is operated by a float. As the fuel level changes, the float adjusts the potentiometer and its resistance changes. The fuel gage is nothing more than an ohmmeter that indicates the resistance of the fuel level sensor.

For discrete sensors, there are two types of outputs, the **NPN** or **sinking output** and the **PNP** or **sourcing output.** The NPN or sinking output has an output circuit that functions similar to a TTL open collector output. It can be regarded as an NPN bipolar transistor with a grounded emitter and an uncommitted collector, as shown in Figure 8–2. In reality, this output circuit could be composed of an actual NPN transistor, an FET, an opto-isolator, or even a relay or switch contact. However, no matter how the output circuitry is composed, in operation it presents either an open circuit or a grounded line for its two output logical signals.

FIGURE 8–2

Sensor with NPN Output

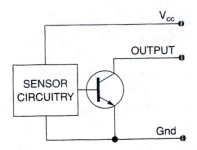

Although this may seem rather convoluted and confusing at first, there is a specific application for an output circuit such as this. Since one of the logical states of the output is an open circuit, it can be used to drive loads that are powered by voltages higher than the power supply range of the sensor. This means that it is capable of operating a load that is being powered from a separate power supply, as shown in Figure 8–3. For example, it is relatively easy to have a sensor that operates from a +10 VDC power supply (V_{sensor}) operate a load that is powered by a +24 VDC (V_{load}). Of course, it is also permissible to operate the NPN output from the same power supply. Typically, sensors with NPN outputs are capable of controlling load voltages up to 30 VDC. The power supply for the load can be any voltage between zero and the maximum collector-to-emitter voltage ($V_{CE(br)}$) specified for the output transistor.

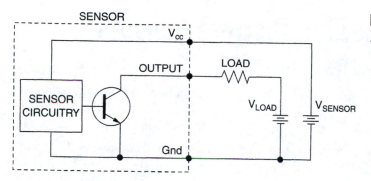

FIGURE 8–3

NPN Sensor Load Connection

The PNP, or sourcing output, has output logic levels that switch between the sensor's power supply voltage and an open circuit. In this case, as illustrated in Figure 8–4, the PNP output transistor has the emitter connected to V_{cc} and the collector uncommitted. When the output is connected to a grounded load, the transistor will cause the load voltage to be either zero (when the transistor is off) or approximately V_{cc} (when the transistor is on).

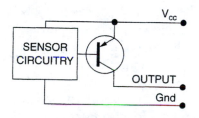

FIGURE 8–4

Sensor with PNP Output

This is ideal for supplying loads that have power supply requirements that are the same as that of the sensor, and when one of the two connection wires of the load is already connected to ground. Notice in Figure 8–5 that this allows a simpler design because only one power supply is needed. However, the disadvantage in this type of circuit is that the sensor and the load must be selected so that they operate from the same supply voltage.

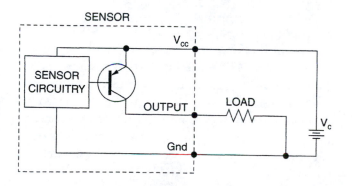

FIGURE 8–5

PNP Sensor Load Connection

8-2 Connecting Discrete Sensors to PLC Inputs

Since discrete PLC inputs can be either sourcing or sinking, it is important to know how to select the sensor output type that will properly interface with the PLC input, and how to wire the PLC input so that it will interface properly to the sensor. Generally speaking, *sensors with sourcing (PNP) outputs should be connected to sinking PLC inputs*, and *sensors with sinking (NPN) outputs should be connected to sourcing PLC inputs*. Connecting sourcing sensor outputs to sourcing PLC inputs, or sinking outputs to sinking inputs, will result in erratic, illogical operation at best, or most likely, a system that will not function at all.

For sourcing (PNP) sensor outputs, the PLC input circuit is wired with the common terminal connected to the common of the sensor as shown in Figure 8–6. When the PNP transistor in the sensor is off, no current flows between the sensor and the PLC, and the PLC input will be OFF. When the sensor circuitry switches the PNP transistor ON, current flows from the V_{cc} power supply, through the PNP transistor, through the IN0 opto-isolator in the PLC input, and out of the common terminal to return to the negative side of the power supply. In this case, the PLC input will be ON. For this type of connection, the value of the V_{cc} voltage must be at least high enough to satisfy the minimum input voltage requirement for the PLC inputs. Notice the simplicity of the connection scheme. The sensor connects directly to the PLC input. No other external signal conditioning circuitry is required.

We can also connect a sinking (NPN) sensor output to the same PLC input. However, since the sensor has a sinking output, the PLC must be rewired as a sourcing input. This can be done by disconnecting the common terminal of the PLC input from the negative side of V_{cc} and connecting it to the positive side of V_{cc} as shown in Figure 8–7. This connection scheme converts all of the PLC inputs to sourcing; that is, in order to switch the PLC input ON, we must draw current out of the input terminal. In operation, when the NPN transistor in the sensor is OFF, no current flows between the sensor and PLC. However, when the NPN transistor switches ON, current will flow from the positive side of the V_{cc} supply, into the common terminal of the PLC, up through the opto-isolator, out of the PLC input terminal IN0, and through the NPN transistor to ground. This will switch the PLC input ON. If it is necessary to operate the PLC inputs and the sensor from separate power supplies, it is permissible as long as the negative terminal of both power supplies are connected together.

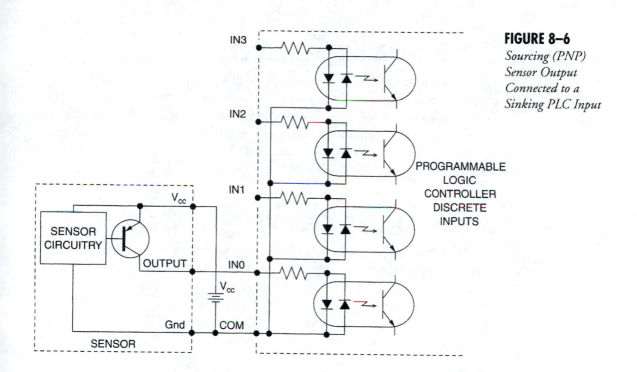

FIGURE 8–6

Sourcing (PNP) Sensor Output Connected to a Sinking PLC Input

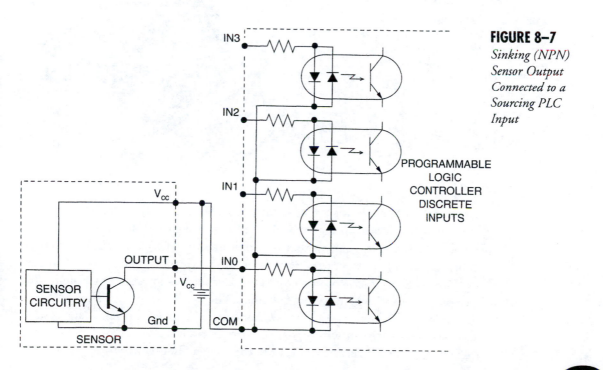

FIGURE 8–7

Sinking (NPN) Sensor Output Connected to a Sourcing PLC Input

8-3 Proximity Sensors

Proximity sensors are discrete sensors that sense when an object has come near to the sensor face. There are four fundamental types of proximity sensors—the **inductive proximity sensor,** the **capacitive proximity sensor,** the **ultrasonic proximity sensor,** and the **optical proximity sensor.** In order to properly specify and apply proximity sensors, it is important to understand how they operate and to which applications each is best suited.

8-4 Inductive Proximity Sensors

As with all proximity sensors, inductive proximity sensors are available in various sizes and shapes as shown in Figure 8–8. As the name implies, inductive proximity sensors operate on the principle that the inductance of a coil and the power losses in the coil vary as a metallic (or conductive) object is passed near it. Because of this operating principle, inductive proximity sensors are only used for sensing metal objects. They will not work with non-metallic materials.

FIGURE 8–8

Samples of Inductive Proximity Sensors (Courtesy of Pepperl & Fuchs, Inc.)

To understand how inductive proximity sensors operate, consider the cutaway block diagram shown in Figure 8–9. Mounted just inside the face of the sensor (on the left end) is a coil that is part of the tuned circuit of an oscillator. When the oscillator operates, there is an alternating magnetic field (called a sensing field) produced by the coil. This magnetic field radiates through the face of the sensor (which is non-metallic). The oscillator circuit is tuned such that as long as the sensing field senses non-metallic material (such as air) it will continue to oscillate, it will trigger the trigger circuit, and the output switching device (which inverts the output of the trigger circuit) will be off. The sensor will therefore send an OFF signal through the cable extending from the right side of the sensor in Figure 8–9.

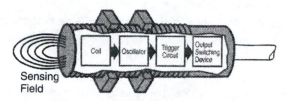

FIGURE 8–9

Inductive Proximity Sensor Internal Components (Courtesy of Pepperl & Fuchs, Inc.)

When a metallic object (steel, iron, aluminum, tin, copper, etc.) comes near the face of the sensor, as shown in Figure 8–10, the alternating magnetic field in the target produces circulating eddy currents inside the material. To the oscillator, these eddy currents are a power loss. As the target moves nearer, the eddy current loss increases, which loads the output of oscillator. This loading effect causes the output amplitude of the oscillator to decrease.

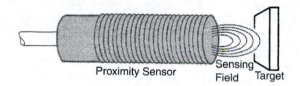

FIGURE 8–10

Inductive Proximity Sensor Sensing Target Object (Courtesy of Pepperl & Fuchs, Inc.)

As long as the oscillator amplitude does not drop below the threshold level of the trigger circuit, the output of the sensor will remain off. However, as shown in Figure 8–11, if the target object moves closer to the face of the sensor, the eddy current loading will cause the oscillator to stall (cease to oscillate). When this happens, the trigger circuit senses the loss of oscillator output and causes the output switching device to switch ON.

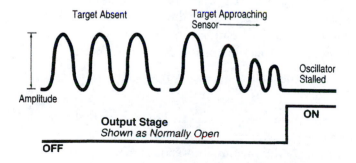

FIGURE 8–11

Inductive Proximity Sensor Signals (Courtesy of Pepperl & Fuchs, Inc.)

The sensing range of a proximity sensor is the maximum distance the target object may be from the face of the sensor in order for the sensor to detect it. One parameter affecting the sensing range is the size (diameter) of the sensing coil. Small diameter sensors (approximately ¼" in diameter) have typical sensing ranges in the area of 1 mm, while large diameter sensors (approximately 3" in diameter) have sensing ranges in the order of 50 mm or more. Additionally, since different metals have different values of resistivity (which limits the eddy currents) and permeability (which channels the magnetic field through the target), the type of metal being sensed will affect the sensing range. Inductive proximity sensor manufacturers derate their sensors based on different metals, with steel being the reference (i.e., having a derating factor of 1.0). Some other approximate derating factors are stainless steel, 0.85; aluminum, 0.40; brass, 0.40; and copper, 0.30.

As an example of how to apply the derating factors, assume you are constructing a machine to automatically count copper pennies as they travel down a chute, and the sensing distance will be 5 mm. In order to detect copper (derating factor 0.30), you would need to purchase a sensor with a sensing range of 5 mm / 0.30 = 16.7 mm. Let's say you found a sensor in stock that has a sensing range of 10 mm. If you use this to sense the copper pennies, you would need to mount it near the chute so that the pennies will pass within (10 mm)(0.30) = 3 mm of the face of the sensor.

Inductive proximity sensors are available in both dc- and ac-powered models. Most require three electrical connections—ground, power, and output. However, there are other variations that require two wires and four wires. Most sensors are available with a built-in LED that indicates when the sensor output is on. One of the first steps a designer should take when using any proximity sensor is to acquire a manufacturer's catalog and investigate the various types, shapes, and output configurations to determine the best choice for the application.

Since the parts of machines are generally constructed of some type of metal, there is an enormous number of possible applications for inductive proximity sensors. They are relatively inexpensive (approximately $30 and up), extremely reliable, operate from a wide range of power supply voltages, are rugged, and because they are totally self contained, they connect directly to the discrete inputs on a PLC with no additional external components. In many cases, inductive proximity sensors make excellent replacements for mechanical limit switches.

To illustrate some of the many possible applications of inductive proximity sensors (sometimes called **inductive prox**), consider these uses:

● By placing an inductive prox next to a gear, the prox can sense the passing gear teeth to give rotating speed information. This application is currently used as a speed feedback device in automotive cruise control systems where the prox is mounted in the transmission.

- All helicopters have an inductive prox mounted in the bottom of the rotor gearbox. Should the gears in the gearbox shed any metal chips (indicating an impending catastrophic failure of the gearbox), the inductive prox senses these chips and activates a warning light on the cockpit instrument panel.

- Inductive proxes can be mounted on access doors and panels of machines. The PLC can be programmed to shut down the machine if any of these doors and access panels are opened.

- Very large inductive proxes can be mounted in roadbeds to sense passing automobiles. This technique is currently used to operate traffic lights.

8-5 Capacitive Proximity Sensors

Capacitive proximity sensors are available in shapes and sizes similar to the inductive proximity sensor (as shown in Figure 8–12). However, because of the principle upon which the capacitive proximity sensor operates, applications for the capacitive sensor are somewhat different.

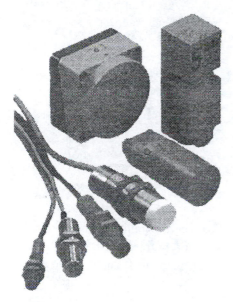

FIGURE 8–12

Example of Capacitive Proximity Sensors (Courtesy of Pepperl & Fuchs, Inc.)

To understand how capacitive proximity sensors operate, consider the cutaway block diagram shown in Figure 8–13. The principle of operation of the sensor is that an internal oscillator will not oscillate until a target material is moved close to the sensor face. The target material varies the capacitance of a capacitor in the face of the sensor that is part of the oscillator circuit. There are two types of capacitive sensor, each with a different way that this sensing capacitor is formed. In the **dielectric type capacitive proximity sensor,** there are two side-by-side capacitor plates in the sensor face. For this type of sensor, the external target acts as the dielectric. As the target is moved closer to the sensor face, the change in dielectric increases the capacitance of the internal capacitor, making the oscillator amplitude increase, which in turn causes the sensor to output an ON signal. The **conductive type capacitive proximity sensor** operates similarly; however, there is only one capacitor plate in the sensor face. The target becomes the other plate. Therefore, for this type of sensor, it is best if the target is an electrically conductive material (usually metal or water-based).

FIGURE 8–13

*Capacitive
Proximity Sensor
Internal
Components
(Courtesy of Pepperl
& Fuchs, Inc.)*

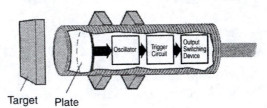

Target   Plate

As is shown in the waveform diagram in Figure 8–14, as the target approaches the face of the sensor, the oscillator amplitude increases, which causes the sensor output to switch on.

FIGURE 8–14

*Capacitive
Proximity Sensor
Signals
(Courtesy of Pepperl
& Fuchs, Inc.)*

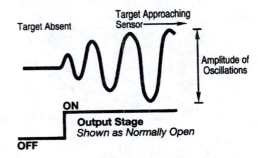

Dielectric capacitive proximity sensors will sense both metallic and non-metallic objects. However, in order for the sensor to work properly, it is best if the material being sensed has a high density. Low-density materials (foam, bubble wrap, paper, etc.) do not cause a detectable change in the dielectric and consequently will not trigger the sensor.

Conductive capacitive proximity sensors require that the material being sensed be an electrical conductor. These are ideally suited for sensing metals and conductive liquids. For example, since most disposable liquid containers are made of plastic or cardboard, these sensors have the unique ability to "look" through the container and sense the liquid inside. Therefore, they are ideal for liquid-level sensors.

Capacitive proximity sensors will sense metal objects just as inductive sensors will. However, capacitive sensors are much more expensive than the inductive types. Therefore, if the material to be sensed is metal, the inductive sensor is the more economical choice.

Some of the potential applications for capacitive proximity sensors include:

- They can be used as a non-contact, liquid-level sensor. They can be placed outside a container to sense the liquid level inside. This is ideal for milk, juice, or soda bottling operations.

- Capacitive proximity sensors can be used as replacements for pushbuttons and palm switches. They will sense the hand and, because they have no moving parts, they are more reliable than mechanical switches.

- Since they are hermetically sealed, they can be mounted inside liquid tanks to sense the tank-fill level.

As with the inductive proximity sensors, capacitive proximity sensors are available with a built-in LED indicator to indicate the output logical state. Also, because capacitive proximity sensors are used to sense materials with a wide range of densities, manufacturers usually provide a sensitivity adjusting screw on the back of the sensor. Then, when the sensor is installed, the sensitivity can be adjusted for best performance in the particular application.

8-6 Ultrasonic Proximity Sensors

The ultrasonic proximity sensor operates using the same principle as shipboard sonar. As shown in Figure 8–15, an ultrasonic "ping" is sent from the face of the sensor. If a target is located in front of the sensor and is within range, the ping will be reflected by the target and returned to the sensor. When an echo is returned, the sensor detects that a target is present, and by measuring the time delay between the transmitted ping and the returned echo, the sensor can calculate the distance between the sensor and the target.

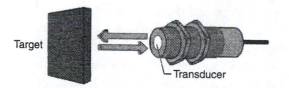

FIGURE 8–15

Ultrasonic Proximity Sensor (Courtesy of Pepperl & Fuchs, Inc.)

As with any proximity sensor, the ultrasonic prox has limitations. The sensor is only capable of sensing a target that is within the sensing range. The sensing range is a funnel-shaped area directly in front of the sensor, as shown in Figure 8–16. Sound waves travel from the face of the sensor in a cone-shaped dispersion pattern bounded by the sensor's **beam angle.** However, because the sending and receiving transducers are both located in the face of the sensor, the receiving transducer is "blinded" for a short period of time immediately after the ping is transmitted—similar to the way our eyes are temporarily blinded by a flashbulb. This means that any echo that occurs during this "blind" time period will go undetected. These echos will be from targets that are very close to the sensor within what is called the sensor's **deadband.** In addition, because of the finite sensitivity of the receiving transducer, there is a distance beyond which the returning echo cannot be detected. This is the **maximum range** of the sensor. These constraints define the sensor's **useable sensing area.**

FIGURE 8–16

Ultrasonic Proximity Sensor Useable Sensing Area (Courtesy of Pepperl & Fuchs, Inc.)

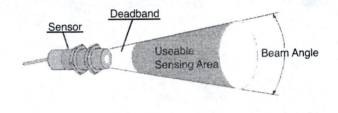

Ultrasonic proximity sensors that have a discrete output generally have a switch-point adjustment provided on the sensor that allows the user to set the target distance at which the sensor output switches on. Note that ultrasonic sensors are also available with analog outputs that will provide an analog signal proportional to the target distance. These types are discussed later in this chapter in the section on distance sensors.

Ultrasonic proximity sensors are useful for sensing targets that are beyond the very short operating ranges of inductive and capacitive proximity sensors. Off the shelf, ultrasonic proxes are available with sensing ranges of 6 meters or more. They sense dense target materials such as metals and liquids best. They do not work well with soft materials, such as cloth, foam rubber, or any material that is a good absorber of sound waves, and they operate poorly with liquids that have surface ripple or waves. Also, for obvious reasons, these sensors will not operate in a vacuum. Since the sound waves must pass through the air, the accuracy of these sensors is subject to the sound propagation time of the air. The most detrimental impact of this is that the sound propagation time of air decreases by 0.17%/° C. This means that as the air temperature increases, a stationary target will appear to move closer to the sensor. They are not affected by ambient audio noise, nor by wind. However, because of their relatively long useful range, the system designer must take care when using more than one ultrasonic sensor in a system because of the potential for crosstalk between sensors.

One popular use for the ultrasonic proximity sensor is in sensing liquid level. Figure 8–17 shows such an application. Note that since ultrasonic sensors do not perform well with liquids with surface turbulence, a stilling tube is used to reduce the potential turbulence on the surface of the liquid.

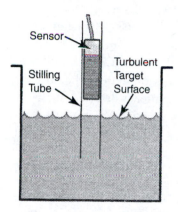

FIGURE 8–17

Ultrasonic Liquid Level Sensor (Courtesy of Pepperl & Fuchs, Inc.)

8-7 Optical Proximity Sensors

Optical sensors are an extremely popular method of providing discrete-output sensing of objects. Since the sensing method uses light, they are capable of sensing any objects that are opaque, regardless of the color or reflectivity of the surface. They operate over long distances (as opposed to inductive or capacitive proximity sensors), will sense in a vacuum (as opposed to ultrasonic sensors), and can sense any type of material whether it is metallic, conductive, or porous. Since the optical transmitters and receivers use focused beams (using lenses), they can be operated in close proximity of other optical sensors without crosstalk or interference.

There are fundamentally three types of optical sensors. These are the **thru-beam, diffuse reflective,** and **retro-reflective.** All three types have discrete outputs. These are generally available in one of three types of light source—incandescent light, red LED, and infrared LED. The red LED and IR LED sensors generally have a light output that is pulsed at a high frequency and a receiver that is tuned to the frequency of the source. By doing so, these types have a high degree of immunity to other potentially interfering light sources. Therefore, red LED and IR LED sensors function better than incandescent sensors in areas where there is a high level of ambient light (such as sunlight) or light noise (such as welding). In addition to specifying the sensor type and light source type, the designer also needs to specify whether the sensor output will be on or off when no light is received. Generally, this is specified as the logical condition when there is no light received (i.e., the dark condition). For this reason, the choices are specified as **dark-on** and **dark-off.**

Thru-Beam (Interrupted)

The thru-beam optical sensor consists of two separate units, each mounted on opposite sides of the object to be sensed. As shown in Figure 8–18, one unit (the emitter) is the light source that provides a lens-focused beam of light that is aimed at the receiver. The other unit, the receiver, also contains a focusing lens and is aimed at the light source. Assuming this is a dark-on sensor, when there is nothing blocking the light beam, the light from the source is detected by the receiver, and there is no output from the receiver. However, if an object passes between the emitter and receiver, the light beam is blocked and the receiver switches on its output.

FIGURE 8–18

Thru-Beam Optical Sensor, Dark-On (Courtesy of Pepperl & Fuchs, Inc.)

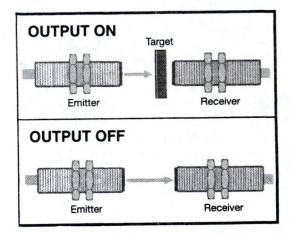

Thru-beam sensors are the most common type known to the general public since they appear in action movies in which thieves are attempting to thwart a matrix of optical burglar alarm sensors setup around a valuable museum piece.

Thru-beam opto sensors work well as long as the object to be sensed is not transparent. They have an excellent (long) maximum operating range. The main disadvantage with this type of sensor is that because the emitter and receiver are separate units, this type of sensor system requires wiring on both sides of the transport system (generally a conveyor) that is moving the target object. In some cases, this may be either inconvenient or impossible. When this occurs, another type of optical sensor should be considered.

Diffuse Reflective (Proximity)

The diffuse reflective optical sensor, shown in Figure 8–19, has the light emitter and receiver located in the same unit. Assuming it is a dark-off sensor, light from the emitter is reflected from the target object being sensed and returned to the receiver, which, in turn, switches on its output. When a target object is not present, no light is reflected to the receiver and the sensor output switches off (dark-off).

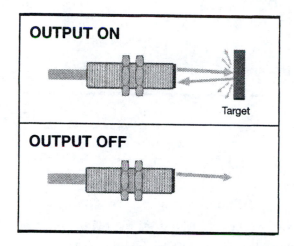

FIGURE 8–19
Diffuse Reflective Optical Sensor, Dark-Off (Courtesy of Pepperl & Fuchs, Inc.)

Diffuse reflective optical sensors are more convenient than thru-beam sensors because the emitter and receiver are located in the same housing, which simplifies wiring. However, this type of sensor does not work well with transparent targets or targets that have a low reflectivity (dull finish, black surface, etc.). Care must also be taken with glossy target objects that have multifaceted surfaces (e.g., automobile wheel covers or corrugated roofing material) or objects that have gaps through which light can pass (e.g., toy cars with windows, compact discs). These types of target objects can cause optical sensors to output multiple pulses for each object.

Retro-Reflective (Reflex)

The retro-reflective optical sensor is the most sophisticated of all of the sensors. Like the diffuse reflective sensor, this type has both the emitter and receiver housed in one unit. As shown in Figure 8–20, the sensor works similar to the thru-beam sensor in that a target object passing in front of the sensor blocks the light being received. However, in this case, the light being blocked is light returning from a reflector. Therefore, this sensor does not require the additional wiring for the remotely located receiver unit.

FIGURE 8–20

Retro-Reflective Optical Sensor, Dark-On (Courtesy of Pepperl & Fuchs, Inc.)

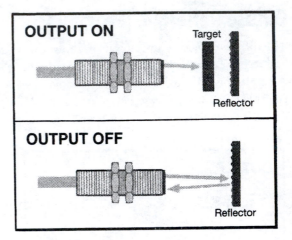

Generally, this type of sensor would not work well with glossy target objects because they would reflect light back to the receiver just as the remote reflector would. However, this problem is avoided by using polarizing filters. This polarizing filter scheme is illustrated in Figure 8–21. Notice in our illustration that there is an added polarizing filter that polarizes the exiting light beam. In our illustration, this is a horizontal polarization. In Figure 8–21a, notice that, with no target object present, the specially designed reflector twists the polarization angle by 90° and sends the light back in vertical polarization. At the receiver, there is another polarizing filter; however, this filter is installed with a vertical polarization to allow the light returning from the reflector to pass through and be detected by the receiver.

In Figure 8–21b, notice that when a target object passes between the sensor and the reflector, not only is the light beam disrupted, but if the object has a glossy surface and reflects the light beam, the reflected beam returns with the same horizontal polarization as the emitted beam. Since the receiver filter has a vertical polarization, the receiver does not receive the light so it activates its output.

Retro-reflective sensors work well with all types of target objects. However, when purchasing the sensor, it is important to also purchase the reflector specified by the manufacturer. These sensors have a maximum range that is more than the diffuse reflective sensor, but less than the thru-beam sensor.

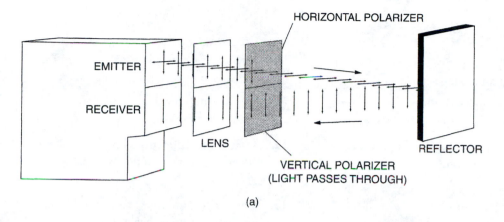

(a)

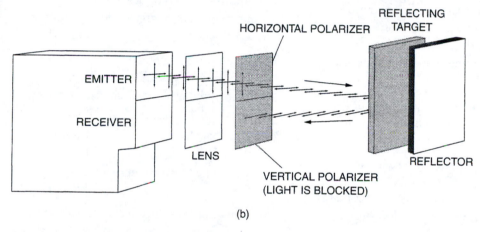

(b)

FIGURE 8–21

Retro-Reflective Optical Sensor Using Polarizing Filters (Exploded View)

Summary

Although we have seen the most commonly used discrete position sensors in this chapter, advanced technologies are providing us with newer types of sensors that are more sensitive, have wider applications, and, in some cases, are intelligent. For example, the use of video and audio equipment is providing sensors with the capability to perform go/no-go image and sound recognition. These types of sensing equipment can perform automated quality inspection or can assist a machine operator in assuring that a machine is operating properly. The growing base of available sensor types is continuously making the system designer's task easier, while at the same time giving the designer more capabilities to design an efficient and reliable control system.

Review Questions

1. What is the difference between a discrete sensor and an analog sensor?

2. Explain the difference between sensors with an NPN output and a PNP output.

3. Name one advantage in using a sensor with an NPN output.

4. Name one advantage in using a sensor with a PNP output.

5. Although inductive proximity sensors operate using the principle of electromagnetic induction, they are capable of sensing non-ferrous metals. Explain why.

6. Manufacturers specify the range of inductive proximity sensors based on what type of target metal?

7. An inductive proximity sensor has a specified range of 5 mm. What is its range when sensing a target object that is a) brass, b) copper, c) stainless steel, and d) aluminum?

8. An inductive proximity sensor is needed that will sense a stainless steel target object that is 7.5 mm from the face of the sensor. We should therefore specify a sensor with what minimum range?

9. In addition to sensing non-metallic objects, capacitive proximity sensors will sense metal objects as well as inductive proximity sensors. So why use inductive proximity sensors?

10. Why aren't ultrasonic proximity sensors used to sense close-range objects?

11. State one disadvantage in using a thru-beam optical sensor.

12. What is the difference between a diffuse reflective and retro-reflective optical sensor?

13. When the beam of a thru-beam, NPN-output, dark-on optical sensor is interrupted by a target object, its output signal will be _____ (high or low).

14. When the beam of a diffuse reflective, PNP-output, dark-off optical sensor is interrupted by a target object, its output signal will be _____ (high or low).

15. When the beam of a retro-reflective, NPN-output, dark-off optical sensor is interrupted by a target object, its output signal will be _____ (high or low).

16. Why do retro-reflective optical sensors use polarizing light filters?

17. Retro-reflective optical sensors require the use of a special reflector. Why is it that a simple glass mirror will not work as the reflector?

Encoders, Transducers, and Advanced Sensors

Objectives

Upon completion of this chapter, you will know:

- the difference between a sensor, a transducer, and an encoder.
- various types of devices to sense and measure temperature, liquid level, force, pressure and vacuum, flow, inclination, acceleration, angular position, and linear displacement.
- how to select a sensor for an application.
- the limitations of each of the sensor types.
- how analog sensor outputs are scaled.

Introduction

In addition to simple, discrete output proximity sensors discussed in the previous chapter, the control system designer also has available a wide variety of sensors that can monitor parameters such as temperature, liquid level, force, pressure and vacuum, flow, inclination, acceleration, position, and others. These types of sensors are usually available with either discrete or analog outputs. If discrete output is available, in many cases the sensors will have a setpoint control so the designer can adjust the discrete output to switch states at a prescribed value of the measured parameter. It should be noted that sensor technology is a rapidly evolving field. Therefore, controls system designers should keep a good source of manufacturers' data and always scan new manufacturers' catalogs to keep abreast of the latest available developments.

This chapter deals with three types of devices—the encoder, the transducer, and the sensor.

1. The **encoder** is a device that senses a physical parameter and converts it to a digital code. In a strict sense, an analog-to-digital converter is an encoder because it converts a voltage or current to a binary coded value.

2. A **transducer** converts one physical parameter into another. The fuel level sending unit in an automobile fuel tank is a transducer because it converts a liquid level to a variable resistance, voltage, or current that can be indicated by the fuel gage.

3. As we saw in the previous chapter, a sensor is a device that senses a physical parameter and provides a discrete, one-bit binary output that switches state whenever the parameter exceeds the setpoint.

9-1 Temperature

There are a large variety of methods for sensing and measuring temperature, from a simple home heating/air conditioning thermostat up to some that require rather sophisticated electronics signal conditioning. This text will not attempt to cover all of the types, but instead will focus on the most popular.

Bi-Metallic Switch

The bi-metallic switch is a discrete (ON-OFF) sensor that takes advantage of the fact that as materials are heated they expand and that for the same change in temperature, different types of material expand differently. As shown in Figure 9–1, the switch is constructed of a bi-metallic strip. The bi-metallic strip consists of two different metals that are bonded together. The metals are chosen so that their coefficients of temperature expansion are radically different. Since the two metals in the strip will be at the same temperature, as the temperature increases, the metal with the larger of the two coefficients of expansion will expand more and cause the strip to warp. If we use the strip as a conductor and arrange it with contacts as shown in Figure 9–1, we will have a bi-metallic switch. Therefore, the bi-metallic strip acts as a relay that is actuated by temperature instead of magnetism.

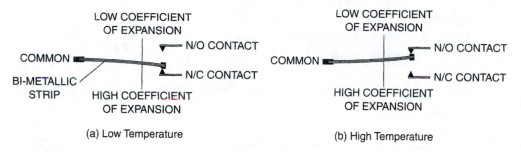

FIGURE 9–1

Bi-metallic Temperature Switch

In most bi-metallic switches, a spring mechanism is added to give the switch a snap action. This forces the strip to quickly snap between its two positions, which prevents arcing and pitting of the contacts as the bi-metallic strip begins to move between contacts. As illustrated in Figure 9–2, the snap action spring is positioned so that no matter which position the bi-metallic strip is in, the spring tends to apply pressure to the strip to hold it in that position. This gives the switch hysteresis (or deadband). Therefore, the temperature at which the bi-metallic strip switches in one direction is different than the temperature that causes it to return to its original position.

FIGURE 9–2

Bi-metallic Temperature Switch with Snap Action

The N/O and N/C electrical symbols for the temperature switch are shown in Figure 9–3. Although the zig-zag line connected to the switch arm symbolizes a bi-metallic strip, this symbol is also used for any type of discrete output temperature switch, no matter how the temperature is sensed. Generally, temperature switches are drawn in the state they would take at room temperature. Therefore, a N/O temperature switch, as shown in Figure 9–3a, would close at some temperature higher than room temperature, and the N/C switch in Figure 9–3b would open at high temperatures. Also, if the switch actuates at a fixed temperature (called the **setpoint**), we usually write the temperature next to the switch, as shown next to the N/C switch in Figure 9–3b. This switch would open at 255° F.

FIGURE 9–3
*Discrete Output
Temperature
Switch Symbols*

(a)   (b)

Thermocouple

Thermocouples provide analog temperature information. They are extremely simple, rugged, repeatable, and very accurate. The operation of the thermocouple is based on the physical property that whenever two different (called **dissimilar**) metals are connected (usually welded), they produce a voltage. The magnitude of the voltage (called the Seebeck voltage) is directly proportional to the temperature of the junction. For certain pairs of dissimilar metals, the temperature-voltage relationship is linear over a small range, however, over the full range of the thermocouple, linearization requires a complex polynomial calculation.

The temperature range of a thermocouple depends only on the two types of dissimilar metals used to make the thermocouple junction. There are six types of thermocouples that are in commercial use, each designated by a letter. These are listed in Table 9–1. Of these, the types J, K, and T are the most popular.

Table 9–1 Popular Thermocouple Types

Type	Metals Used	Temperature Range
E	Chromel-Constantan	−100° C to +1000° C
J	Iron-Constantan	0° C to +760° C
K	Chromel-Alumel	0° C to +1370° C
R	Platinum-Platinum/13%Rhodium	0° C to +1000° C
S	Platinum-Platinum/10%Rhodium	0° C to +1750° C
T	Copper-Constantan	−60° C to +400° C

In Table 9–1, some of the metals are alloys. For example, chromel is a chrome-nickel alloy, alumel is an aluminum-nickel alloy, and constantan is a copper-nickel alloy.

It is important to remember that each time two dissimilar metals are joined, a Seebeck voltage is produced. This means that thermocouples must be wired using special wire that is of the same two metal types as the thermocouple junction to which they are connected. Wiring a thermocouple with off-the-shelf copper hookup wire will create additional junctions and accompanying Seebeck voltages and temperature measurement errors. For example, if we wish to use a type-J thermocouple, we must also purchase type-J wire to use with it, connecting the iron wire to the iron side of the thermocouple and the constantan wire to the constantan side of the thermocouple.

It is not possible to connect a thermocouple directly to the analog input of a PLC or other controller because the Seebeck voltage is too low (generally less than 50 millivolts for all types) to be measured by conventional analog input systems. In addition, since the thermocouples are non-linear over their full range, compensation must be added to linearize their output. Therefore, to go with each type of thermocouple, most manufacturers also market electronic devices that will amplify, condition, and linearize the thermocouple output. As an alternative, most PLC manufacturers offer thermocouple input modules designed for the direct connection of thermocouples. These modules internally provide the signal conditioning needed for the thermocouple type being used.

As with all analog inputs, thermocouple inputs are sensitive to electromagnetic interference, especially since the voltages and currents are extremely low. Therefore, the control system designer must be careful not to route thermocouple wires near or with power conductors. Failure to do so will cause temperature readings to be inaccurate and erratic. In addition, thermocouple wires are never grounded. Each wire pair is always routed all the way to the analog input module without any other intermediate connections.

Resistance Temperature Device (RTD)

All metals exhibit a positive resistance temperature coefficient; that is, as the temperature of the metal rises, so does its ohmic resistance. The resistance temperature device (RTD) takes advantage of this characteristic. The most common metal used in RTDs is platinum because it exhibits better temperature coefficient characteristics and is more rugged than other metals. Platinum has a temperature coefficient of $\alpha = +0.00385$. Therefore, assuming an RTD nominal resistance of 100 ohms at 0° C (one of the typical values for RTDs), its resistance would change at a rate of $+0.385$ ohms/°C. All RTD nominal resistances are specified at 0° C, and the most popular nominal resistance is 100 ohms.

Example 9–1
Problem:

A 100 ohm platinum RTD exhibits a resistance of 123.0 ohms. What is its temperature?

Solution:

Since all RTD nominal resistances are specified at 0° C, the nominal resistance for this RTD is 100 ohms at 0° C, and its change in resistance due to temperature is +23 ohms. Therefore, we divide +23 ohms by 0.385 ohms/° C to get the solution, +59.7° C.

There are two fundamental methods for measuring the resistance of RTDs. First, the Wheatstone bridge can be used. However, keep in mind that since the RTD resistance change is typically large relative to the nominal resistance of the RTD, and since the Wheatstone bridge is linear over a very small range, the voltage output of the Wheatstone bridge will not be a linear representation of the RTD resistance. Therefore, the full Wheatstone bridge equations must be used to calculate the RTD resistance (and corresponding temperature). These equations are available in any fundamental dc circuits text.

The second method for measuring the RTD resistance is the four-wire ohms measurement. This can be done by connecting the RTD to a low current, constant current source with one pair of wires and measuring the voltage drop at the RTD terminals with another pair of wires. In this case, a simple ohms law calculation will yield the RTD resistance. With newer integrated circuit technology, very accurate and inexpensive constant current regulator integrated circuits are readily available for this application.

As with thermocouples, most RTD manufacturers also market RTD signal conditioning circuitry that frees the system designer from this task and makes the measurement system design much easier.

Integrated Circuit Temperature Probes

Temperature probes are now available that contain integrated circuit temperature transducers. These probes generally contain all the required electronics to convert the temperature at the end of the probe to a dc voltage generally between zero and 10 volts dc. These require only a dc power supply input. They are accurate, reliable, extremely simple to apply, and they connect directly to an analog input on a PLC.

9-2 Liquid Level

Float Switch

The liquid level float switch is a simple device that provides a discrete output. As illustrated in Figure 9–4, it consists of a snap-action switch and a long lever arm with a float attached. As the liquid level rises, the lever arm presses on the switch's actuator button. Coarse adjustment of the unit is done by moving the vertical mounting position of the switch. Fine adjustment is done by loosening the mounting screws and tilting the switch slightly (one of the mounting holes in the switch is elongated for this purpose), or by simply bending the lever arm.

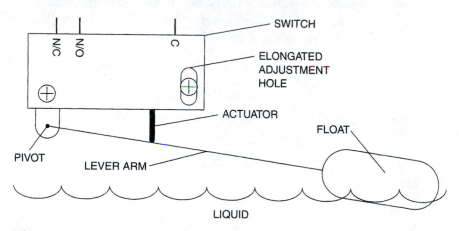

FIGURE 9–4

Liquid Float Switch

The electrical symbols for the float switch are shown in Figure 9–5. The N/O switch in Figure 9–5a closes when the liquid level rises, and the N/C switch in Figure 9–5b opens as the liquid level rises.

FIGURE 9–5

Discrete Output Float Switch Symbols

Float Level Switch

Another variation of the float switch that is more reliable and has fewer moving parts is the float level switch. Although this is not commercially available as a unit, it can be easily constructed. As shown in Figure 9–6, a float is attached to a small section of PVC pipe. A steel ball is dropped into the pipe, and an inductive proximity sensor is threaded and sealed into the pipe. Another section of pipe is threaded to the rear of the sensor, and a pivot point is attached. The point where the sensor wire exits the pipe can be sealed; however, this may not be necessary because sensors are available with the wire sealed where it exits the sensor. When the unit is suspended by the pivot point in an empty tank, the float end will be lower than the pivot point, and the steel ball will be some distance from the prox sensor. However, when the liquid level raises the float so that it is higher than the pivot point, the steel ball will roll toward the sensor and cause the sensor to actuate. In addition, since the steel ball has significant mass when compared to the entire unit, when the ball rolls to the sensor the center of gravity will shift to the left causing the float to rise slightly. This will tend to keep the ball to the left, even if there are small ripples on the surface of the liquid. It also means that the liquid level needed to switch on the unit is slightly higher than the level to turn off the unit. This effect is called **hysteresis** or deadband.

FIGURE 9–6

Float Level Switch

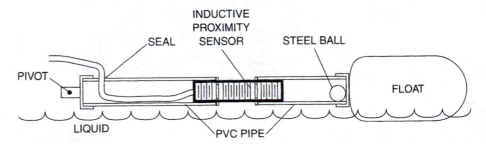

Capacitive

The capacitive proximity sensor used as a liquid level sensor, as discussed in the previous chapter, can provide sensing of liquid level with a discrete output. However, there is another type of capacitive sensor that can provide an analog output proportional to liquid level. This type of sensor requires that the liquid be non-conductive (such as gasoline, oil, alcohol, etc.). For this type of sensor, two conductive electrodes are positioned vertically in the tank so that they are in close proximity, parallel, and at a fixed distance apart. When the tank is empty, the capacitance between the electrodes will be small because the dielectric between them will be air. However, as the tank is filled, the liquid replaces the air dielectric between the probes and the capacitance will increase. The value of the capacitance is proportional to the height of the liquid in the tank.

The capacitance between the electrodes is usually measured using an ac Wheatstone bridge. The differential output voltage from the bridge is then rectified and filtered to produce a dc voltage that is proportional to the liquid level. Any erratic variation in the output of the sensor due to the sloshing of the liquid in the tank can be removed by using either a stilling tube or by low-pass filtering of the electrical signal.

9-3 Force

When a force is applied to a unit area of any material (called **stress**), the material undergoes temporary deformation called **strain.** The strain can be positive (tensile strain) or negative (compression strain). For a given cross-sectional area and stress, most materials have a very predictable and repeatable strain. By knowing these characteristics, we can measure the strain and calculate the amount of stress (force) being applied to the material.

The strain gage is a fundamental building block in many sensors. Since it is capable of indirectly measuring force, it can also be used to measure any force-related unit such as weight, pressure (and vacuum), gravitation, flow, inclination, and acceleration. Therefore, because of this widespread use, it is very important to understand the theory regarding strain and strain gages.

Mathematically, strain is defined as the change in length of a material with respect to the overall length prior to stress, or $\Delta L/L$. Since the numerator and denominator of the expression are in the same units (length), strain is a dimensionless quantity. The lower case Greek letter epsilon, ε, is used to designate strain in mathematical expressions. Because the strain of most solid materials is very small, we generally factor out 10^{-6} from the strain value and then define the strain as MicroStrain, µStrain, or µε.

Example 9–2
Problem:

A 1-foot metal rod is being stretched lengthwise by a given force. Under stress, it is found that the length increases by 0.1 mm. What is the strain?

Solution:

First, make sure all length values are in the same unit of measure. We will convert 1 foot to millimeters. 1 foot = 12 in/ft $\times$ 25.4 mm/in = 304.8 mm/ft. The strain is $\varepsilon = \Delta L/L = 0.1$ mm/304.8 mm = 0.000328, and the microstrain is µε = 328. Since the rod is stretching, ΔL is positive and therefore the strain is positive.

The most common method used for measuring strain is the strain gage. As shown in Figure 9–7, the strain gage is a printed circuit on a thin flexible substrate (sometimes called a **carrier** or **backing**). A majority of the conductor length of the printed circuit is oriented in one direction (in the figure, the orientation is left and right), which is the direction of strain that the strain gage is designed to measure. The strain gage is bonded to the material being measured using a special adhesive that will accurately transmit mechanical strain from the material to the strain gage. When the gage is mounted on the material to be measured and the material is stressed, the strain gage strains (stretches or compresses) with the tested material. This causes a change in the length of the conductor in the strain gage, which causes a corresponding change in its resistance. Note that because of the way the strain gage is designed, it is sensitive to stress in only one direction. In our figure, if the gage is stressed in the vertical direction, it will undergo very little change in resistance. If it is stressed at an oblique angle, it will measure the strain component in one direction only (horizontal, for this figure). If it is desired to measure strain on multiple axes, one strain gage is needed for each axis, with each one mounted in the proper direction corresponding to its particular axis. If two strain gages are sandwiched one on top of another and at right angles to each other, then it is possible to measure oblique strain by vectorially combining the strain readings from the two gages. Strain gages with thinner, longer conductors are more sensitive to strain than those with thick, short conductors. By controlling these characteristics, manufacturers are able to provide strain gages with a variety of sensitivities (called **gage factor**). In strain equations, gage factor is indicated by the variable k. Mathematically the strain gage factor k is the change in resistance divided by the nominal resistance, divided by the strain. The mathematical expression is

$$k = \frac{\Delta R / R}{\varepsilon}$$

where R is the original resistance of the gage, ΔR is the change in resistance of the gage for a given strain, and ε is the strain. Typical gage factors are on the order of one to four.

FIGURE 9–7

Single Axis Strain Gage, Approximately 10 Times Actual Size (Courtesy of Vishay Intertechnology, Inc.)

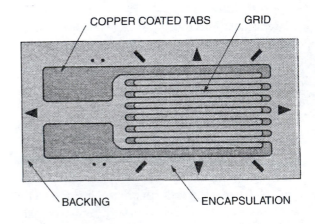

COPPER COATED TABS   GRID

BACKING   ENCAPSULATION

Strain gages are commonly available in nominal resistance values of 120, 350, 600, 700, 1k, 1.5k, and 3k ohms. Because the change in resistance is extremely small with respect to the nominal resistance, strain gage resistance measurements are generally done using a balanced Wheatstone bridge as shown in Figure 9–8. The resistors R_1, R_2, and R_3 are fixed, precision, low temperature coefficient resistors. Resistor R_{STRAIN} is the strain gage. The input to the bridge, V_{in}, is a fixed dc power supply of typically 2.5 to 10 volts. The output V_{out} is the difference voltage from the bridge.

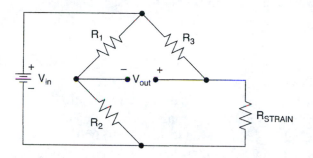

FIGURE 9–8

Strain Gage in a Wheatstone Bridge Circuit

Assuming $R_1 = R_2 = R_3 = R_{STRAIN}$, the bridge will be balanced, and the output V_{out} will be zero. However, should the strain gage be stretched, its resistance will increase slightly and cause V_{out} to increase in the positive direction. Likewise, if the strain gage is compressed, its resistance will be lower than the other three resistors in the bridge, and the output voltage V_{out} will be negative.

For large resistance changes, the Wheatstone bridge is a non-linear device. However, near the balance (zero) point, the bridge is very linear for extremely small output voltages. Since the strain gage resistance change is extremely small, the output of the bridge will likewise be small. Therefore, we can consider the strain gage bridge to be linear. For example, consider the graph shown in Figure 9–9. This is a plot of the Wheatstone bridge output voltage versus strain gage resistance for a 600 ohm bridge (all four resistors are 600 ohms) with $V_{in} = 1$ volt. Notice that even for a strain gage resistance change of 1 ohm (which is unlikely), the bridge output voltage is very linear.

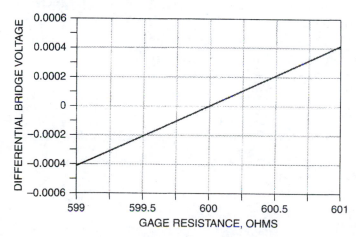

FIGURE 9–9

Wheatstone Bridge Output for 600 Ohm Resistors

Example 9–3
Problem:

A 600 ohm strain gage is connected into a bridge circuit with the other three resistors, $R_1 = R_2 = R_3 = 600$ ohms. The bridge is powered by a 10 volt dc power supply. The strain gage has a gage factor $k = 2.0$. The bridge is balanced ($V_{out} = 0$ volts) when the strain $\mu\varepsilon = 0$. What is the strain when $V_{out} = +500$ microvolts?

Solution:

First, we will find the resistance of the strain gage. Referring back to Figure 9–8, because V_{in} is 10 volts, the voltage at the node between R_1 and R_2 is exactly 5 volts with respect to the negative terminal of the power supply. Therefore, the voltage at the node between R_3 and R_{STRAIN} must be 5 volts + 500 microvolts, or 5.0005 volts. By Kirchoff's voltage law, the voltage drop on R_3 is 10 v − 5.0005 = 4.9995 v. This makes the current through R_3 and R_{STRAIN} 4.9995 v / 600 ohms = 8.333 milliamperes. Therefore, by ohm's law, R_{STRAIN} = 5.0005 v / 8.333 mA = 600.12 ohms. Since the nominal value of R_{STRAIN} is 600 ohms, $\Delta R = 0.12$ ohm.

The gage factor is defined as $k = (\Delta R/R)/(\varepsilon)$. Solving for ε, we get $\varepsilon = (\Delta R/R)/k$. Therefore the strain $\varepsilon = (0.12 / 600) / 2 = 0.0001$, or the microstrain is $\mu\varepsilon = 100$. (*Note that an output of 500 microvolts corresponded to a microstrain of 100. Therefore, since we consider this to be a linear function, we can now define the calibration of this strain gage as $\mu\varepsilon = V_{out}/5$.*)

One fundamental problem associated with strain gage measurements is that generally the strain gage is physically located some distance from the remaining resistors in the bridge. Since the change in resistance of the strain gage is very small, the resistance of the wire between the bridge and strain gage unbalances the bridge and creates a measurement error. The problem is worsened by the temperature coefficient of the wire, which causes the measurement to drift with temperature. (Generally, strain gages are designed to have a very low temperature coefficient, but wire does not.) These problems can be minimized by using what is called a three-wire measurement as shown in Figure 9–10.

FIGURE 9–10
Wheatstone Bridge Circuit with Three-Wire Strain Gage Connection

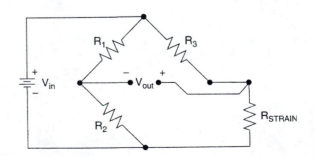

In this circuit, it is crucial for the wire from R_3 to R_{STRAIN} to be identical in length, AWG size, and copper type to the wire from R_{STRAIN} to R_2. By doing so, the voltage drops in the two wires will be equal. If we then add a third wire to measure the voltage on R_{STRAIN}, the resistance of the wire from R_3 to R_{STRAIN} effectively becomes part of R_3, and the resistance of the wire from R_{STRAIN} to R_2 effectively becomes part of R_{STRAIN}, which re-balances the bridge. This physical three-wire strain gage connection scheme is shown in Figure 9–11.

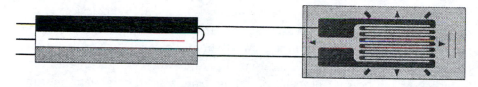

FIGURE 9–11

Three-Wire Connection to Strain Gage (Courtesy of Vishay Intertechnology, Inc.)

9-4 Pressure/Vacuum

Since many machine systems use pneumatic (air) pressure, vacuum, or hydraulic pressure to perform certain tasks, it is often necessary to sense the presence of pressure or vacuum, and in many cases, to measure the magnitude of the pressure or vacuum. Next, we will discuss some of the more popular methods for the discrete detection and the analog sensing of pressure and vacuum.

Bellows Switch

The bellows switch is a relatively simple device that provides a discrete (ON or OFF) signal based on pressure. Referring to Figure 9–12, notice that the bellows

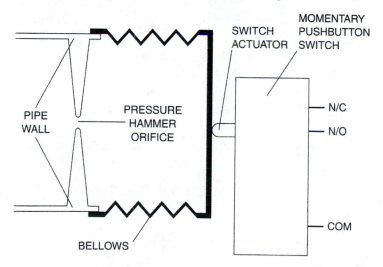

FIGURE 9–12

Bellows-Type Pressure Switch Cross Section

(which is made of a flexible material, usually rubber) is sealed to the end of a pipe from which the pressure is to be sensed. When the pressure in the pipe increases, the bellows pushes on the actuator of a switch. When the pressure increases to a point where the bellows overcomes the switch's actuator spring force, the switch actuates, the N/O contact connects to the common, and the N/C contact disconnects from the common.

Generally, for most pressure sensing switches and sensors, a **pressure hammer** orifice (also called a pressure hammer snubber) is included in the device. This is done to protect the device from extreme pressure transients (called pressure hammer) caused by the closing of valves elsewhere in the system, which could rupture the bellows. Pressure hammer is most familiar to us when we quickly turn off a water faucet and hear the pipes in our homes bang from the transient pressure. The orifice is simply a constriction in the pipe's inner diameter so that air or fluid inside the pipe is prevented from flowing rapidly into the bellows.

The pressure at which the bellows switch actuates is difficult to predict because, in addition to pressure, it also depends on the elasticity of the bellows, the spring force in the switch, and the mechanical friction of the actuator. Therefore, these types of switches are generally used for coarse pressure sensing. The most common use is to simply detect the presence or absence of pressure on the system so that a controller (PLC) can determine if a pump has failed.

Because of the frailty of the rubber bellows, bellows switches cannot be used to sense high pressures. For high-pressure applications, the bellows is replaced by a diaphragm made of a flexible material (nylon, aluminum, etc.) that deforms (bulges) when high pressure is applied. This deformation presses on the actuator of a switch.

The N/O and N/C electrical symbols for the pressure switch are shown in Figure 9–13. The semicircular symbol connected to the switch arm symbolizes a pressure diaphragm. Generally, pressure switches are drawn in the state they would take at 1 atmosphere of pressure (i.e., atmospheric pressure at sea level). Therefore, a N/O pressure switch, as shown in Figure 9–13a, would close at some pressure higher than 0 psig (1 atmosphere), and the N/C switch in Figure 9–13b would open at some pressure higher then 0 psig. Also, if the switch actuates at a fixed pressure, or setpoint, we usually write the setpoint pressure next to the switch as shown next to the N/C switch in Figure 9–13b. This switch would open at 300 psig.

FIGURE 9–13
Discrete Output
Pressure Switch
Symbol

(a)   (b)

300 psi

Strain Gage Pressure Sensor

The strain gage pressure sensor is the most popular method of making analog measurements of pressure. It is relatively simple, reliable, and accurate. It operates on the principle that whenever fluid pressure is applied to any solid material, the material deforms (strains). If we know the strain characteristics of the material and we measure the strain, we can calculate the applied pressure.

For this type of measurement, a different type of strain gage is used, as shown in Figure 9–14. This strain gage measures radial strain instead of longitudinal strain. Notice that there are two different patterns of strain conductors on this strain gage. The pair around the edge of the gage (we will call the "outer gages") appear as regular strain gages but in a curved pattern. Then, there are two semicircular gage patterns in the center of the gage (we will call the "inner gages"). As we will see, each pattern serves a specific purpose in contributing to the pressure measurement.

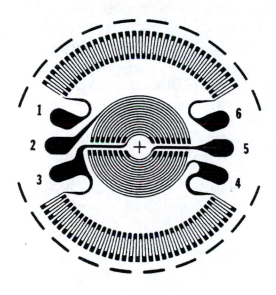

FIGURE 9–14

Diaphragm Strain Gage (Courtesy of Vishay Intertechnology, Inc.)

If we were to glue the diaphragm strain gage to a metal disk with known stress/strain characteristics, and then seal the assembly to the end of a pipe, we would have a unit similar to that in Figure 9–15a. The strain gage and disk assembly are mounted to the pipe with the strain gage on the outside (in Figure 9–15a).

When pressure is applied to the inside of the pipe, the disk and strain gage begin to bulge slightly outward, as shown in Figure 9–15b (an exaggerated illustration). The amount of bulging is proportional to the inside pressure. Referring to Figure 9–15b, notice that the top surface of the disk will be in compression around the outer edge

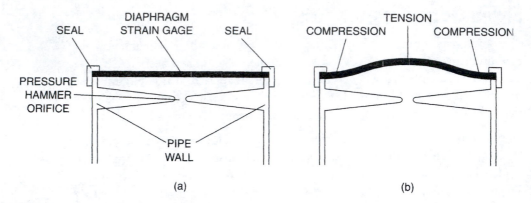

FIGURE 9–15

Strain Gage Pressure Transducer Cross Section (a) 1 Atmosphere, (b) > 1 Atmosphere

(where the outer gages are located) and will be in tension near the center (where the inner gages are located). Therefore, when pressure is applied, the resistance of the outer gages will decrease, and the resistance of the inner gages will increase.

Following the conductor patterns on the diaphragm strain gage in Figure 9–14, if we connect terminals 1 to 2 and 5 to 6, they will be connected in a Wheatstone bridge arrangement. (The reader is encouraged to trace the patterns and draw an electrical diagram.) If we connect the gages as shown in Figure 9–16, we can measure the strain (and therefore the pressure) by measuring the voltage difference V_a–V_b. When pressure is applied, since the resistance of the outer gages will decrease and the resistance of the inner gages will increase, the voltage V_a will increase and the voltage V_b will decrease, which will unbalance the bridge. The voltage difference V_a–V_b will be positive; therefore indicating positive pressure.

FIGURE 9–16

Strain Gage Pressure Transducer Electrical Connections

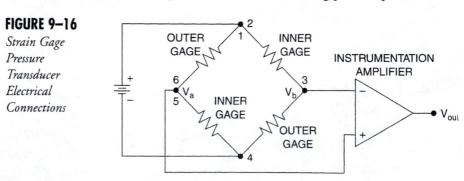

If a vacuum is applied to the sensor, the disk and strain gage will deform (bulge) inward. This causes an exact opposite effect in all the resistance values, which will produce a negative voltage output from the Wheatstone bridge.

Many strain gage type pressure sensors (also called **pressure transducers**) are available with the instrumentation amplifier included inside the sensor housing. The entire unit is calibrated to produce a precise output voltage proportional to pressure. These units have an output that is usually specified as a **calibration factor** in psi/volt.

Example 9–4
Problem:

A 0 to +250 psi pressure transducer has a calibration factor of 25 psi/volt. a) What is the pressure if the transducer output is 2.37 volts? b) What is the full-scale output voltage of the transducer?

Solution:

a. When working a problem of this type, dimensional analysis helps. The calibration factor is in psi/volt, and the output is in volts. If we multiply psi/volt times volts, we get psi—the desired unit for the solution. Therefore, the pressure is 25 psi/volt $\times$ 2.37 v = 59.25 psi.

b. The full scale output voltage is 250 psi / 25 psi/volt = 10 volts.

Variable Reluctance Pressure Sensor

Another method for measuring pressure and vacuum that is more sensitive than most other methods is the variable reluctance pressure sensor (or transducer). This unit operates similarly to the linear variable differential transformer (LVDT) discussed later in this chapter. Consider the cross section illustration of the variable reluctance pressure sensor shown in Figure 9–17. Two identical coils (same dimensions, same number of turns, and same size wire) are suspended on each side of a ferrous diaphragm disk. The disk is sealed to the end of a small tube or pipe. The two coils are connected to an ac Wheatstone bridge with two other matched coils, L3 and L4. If the pressure inside the pipe is 1 atmosphere, the disk will be flat and the inductance of each of the two coils will be the same (L1 = L2). In this case, the bridge will be balanced and V_{out} will be zero. If there is pressure inside the pipe, the diaphragm will deform (bulge) upward. Since the disk is made of a ferrous material, and since it has moved closer to L1, the reluctance in coil L1 will decrease, which will increase its inductance. At the same time, since the disk has moved away from L2, its reluctance will increase, which will decrease its inductance. This will unbalance the bridge and produce an ac output. The magnitude of the output is proportional to the displacement of the disk (the pressure), and the phase of the output with respect to the ac source depends on the direction of displacement (pressure or vacuum).

FIGURE 9–17
Variable Reluctance Pressure Transducer (Cross Section)

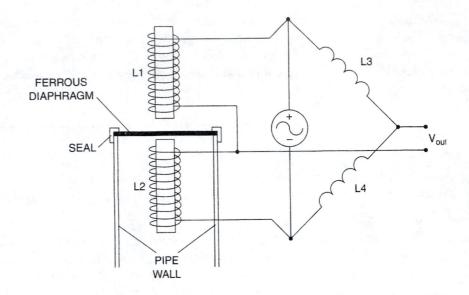

9-5 Flow

The flow, Q, of a fluid in a pipe is directly proportional to the velocity of the fluid, V, and the cross-sectional area of the pipe, A. Therefore, we can say $Q = V \times A$. This is a relatively simple concept to memorize by applying dimensional analysis. For example, in English units, flow is measured in cubic feet per second, velocity in feet per second, and area in square feet, which results in $ft^3/s = ft/s \times ft^2$. Therefore, we may conclude that if we wish to increase the flow of fluid in a pipe, we have two choices: we can increase the fluid's velocity or install a larger diameter pipe. Fluid flow is measured in many different ways—some of which (but not all) are discussed in the following paragraphs. For a comprehensive treatment of fluid-flow measurement techniques, the reader should refer to any manufacturer's tutorial on the subject, such as Omega Engineering, Inc.'s *Fluid Flow and Level Handbook*.

Drag Disk Flow Switch

Any time a moving fluid passes an obstruction, a pressure difference is created with the higher pressure on the upstream side of the obstruction. This pressure difference applies a force to the obstruction that tends to move it in the direction of the fluid flow. The amount of force applied is proportional to the velocity of the fluid (which is proportional to flow rate) and the cross-sectional area the obstruction presents to the flow. For example, a sailboat will move faster if either the wind velocity increases or if we turn the sails so that they present a larger area to the wind. We can take advantage of this force to actuate a switch by using a drag disk as shown in Figure 9–18. This unit consists of a case containing a snap-action switch, a switch lever arm extending from the bottom of the case, and a circular disk attached to the end of the lever arm. In this case, the device is threaded into a "T" connector installed in the pipe and oriented such that the drag disk is perpendicular to the direction of flow. An increase in fluid-flow rate causes a proportional increase in the force on the drag disk. At a predetermined flow rate, the force on the drag disk is sufficient to push the lever arm to the right and actuate the switch.

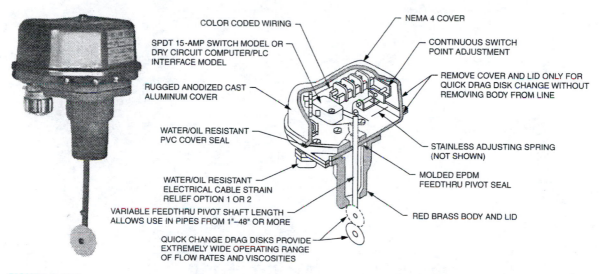

FIGURE 9–18

Drag Disk Flow Switch and Cutaway View
(Courtesy of Harwil Corporation, Patented)

The trip point of this type of switch is adjustable by a screw adjustment that varies the counteracting spring force applied to the lever arm. In addition, the range of adjustment can be changed by changing the cross-sectional area of the drag disk. Switches of this type are usually provided with several drag disks of differing sizes corresponding to various flow ranges.

Another variation of the drag disk flow switch is the in-line flow meter with proximity switch. As shown in Figure 9–19, this device has a drag disk mounted inside a plastic or glass tube with an internal spring to counteract the force applied to the disk by fluid flow. Since the tube is clear and has graduations marked on the outside, the flow rate may also be visually measured by viewing the position of the drag disk inside the tube. A toroidal permanent magnet is mounted on the downstream side of the drag disk and is held in place by the spring force. A magnetic reed switch is mounted on the outside of the tube with its vertical position on the tube being set by an adjustment screw. As fluid flow increases, the disk and piggy-back magnet will rise in the tube. When the magnet aligns with the reed switch, the switch will actuate.

FIGURE 9–19
In-Line Flow Meter with Proximity Switch (Courtesy of PKP GmbH & Clark Solutions)

Thermal Dispersion Flow Switch

A method to indirectly measure flow is to measure the amount of heat that the flow carries away from a heater element that is inserted into the flow. By setting a trip point, we can have an electronic temperature switch actuate when the temperature of the heated probe drops below a predetermined level. This device is called a **thermal dispersion flow switch.** One problem associated with this technique is that, since we may not know the temperature of the fluid in the pipe, it is difficult to determine the flow based simply on the temperature of the heated probe. For example, if we heat the probe to say 50° C, and the fluid temperature happens to be 49° C, then when fluid is moving in the pipe our device will "see" a very slight temperature drop in the heated probe and therefore conclude that the flow is (for all practical purposes) zero. To circumvent this problem, this device actually consists of two probes, as shown in Figure 9–20. One probe is heated, and one is not. The probe that is not heated will measure the static temperature of the fluid, while the heated probe will measure the temperature decrease due to fluid flow. Electronic circuitry then calculates the temperature differential ratio, compares it to the setpoint (which is adjustable using a built-in potentiometer), and outputs a logical ON/OFF signal. One major advantage to this type of temperature switch is that, since it has no moving parts, it is extremely reliable and it works well in dirty fluids that would normally foul the types of probes that have moving parts. Obviously, in dirty fluid systems, the probe must be periodically removed and cleaned in order to keep the setpoint from drifting due to contaminated probes.

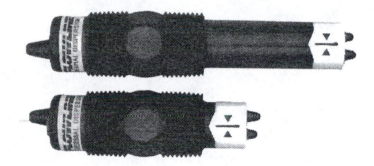

FIGURE 9–20
Thermal Dispersion Flow Switch (Courtesy of Flowline Liquid Intelligence)

Paddle Wheel Flow Sensor

One method to directly measure flow velocity is to simply insert a paddle wheel into the fluid flow and measure the speed at which the paddle wheel rotates. This device, called a **paddle wheel flow sensor** (shown in Figure 9–21), usually provides an output of pulses with the pulse rate being proportional to flow velocity. Most have paddle wheels that are constructed of a material that is impervious to caustic fluids. Since these materials are usually non-metallic, metal beads are embedded in the ends of the paddles so that the proximity sensor will detect them. Paddle wheel flow sensors are most commonly used to meter fuel flow from gas pumps at filling stations. Some of the more elaborate models will also convert the pulse rate to a dc voltage (an analog output). Keep in mind that, as with many types of flow sensors, the device actually measures fluid speed, not flow rate. However, flow rate can be calculated if the pipe diameter is also known.

Turbine Flow Sensor

A variation on the paddle wheel sensor is the **turbine flow sensor,** illustrated in Figure 9–22. In this case, the paddle wheel is replaced by a small turbine that is

FIGURE 9–21
*Paddle Wheel Flow
Sensor*

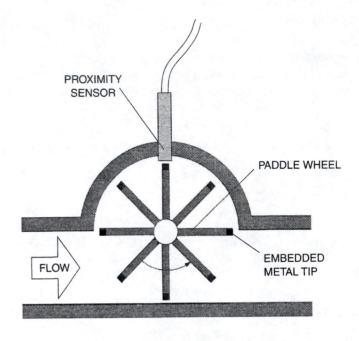

PROXIMITY
SENSOR

PADDLE WHEEL

FLOW

EMBEDDED
METAL TIP

FIGURE 9–22
*Turbine Flow
Sensor, Cutaway
View*

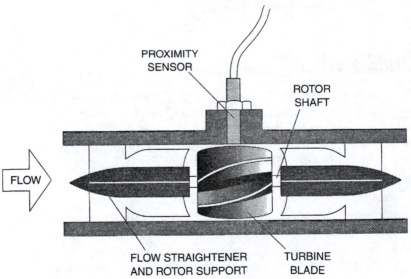

PROXIMITY
SENSOR

ROTOR
SHAFT

FLOW

FLOW STRAIGHTENER
AND ROTOR SUPPORT

TURBINE
BLADE

suspended in the pipe. A special support mechanism routes fluid through the vanes of the turbine without disturbing the flow (i.e., without causing turbulence). The turbine vanes are made of metal (usually brass); therefore, they can be detected by an inductive proximity sensor mounted in the top of the unit. As with the paddle wheel, this device outputs a pulse train with the pulse frequency proportional to the flow velocity.

Pitot Tube Flow Sensor

If a tube is inserted into a fluid flow such that the tip of the tube is aimed in the upstream direction of the flow, the pressure within the tube will be equal to the sum of the **impact pressure** (the pressure caused by the flow) and the **static pressure** of the fluid. It is possible to remove the static pressure from this measurement by separately measuring the static pressure and subtracting it from the overall total pressure. Therefore, the pressure difference (the impact pressure) will be proportional to only the fluid flow. The device that does this is called a **pitot tube.** A pitot tube actually consists of two tubes within a housing. One tube is open at the end of the housing and measures the impact pressure, while the other tube opens at the side of the housing and measures the static pressure. The most common application for the pitot tube is in the sensor for airspeed indicators in aircraft. In this case, the impact pressure is the air pressure created by the velocity of the aircraft, and the static pressure is the atmospheric pressure for the aircraft's altitude. For measuring the flow velocity in a pipe, the principle is the same as that of airspeed indicators, and the pitot tube is constructed similarly. The tube, shown in Figure 9–23, is constructed with two orifices, one to sense the impact pressure and the other for static pressure. The two small pipes contained inside the pitot tube connect these two orifices to two valves on the opposite end of the probe. Two tubes connect the valves on the pitot tube to a differential pressure transducer that has an electrical analog output. In operation, the impact and static pressures are sent from the pitot tube, through the valves, and to the differential pressure transducer. Most manufacturers provide additional electrical circuitry inside the differential pressure transducer that will linearize and calibrate the analog output to be directly proportional to the fluid-flow rate.

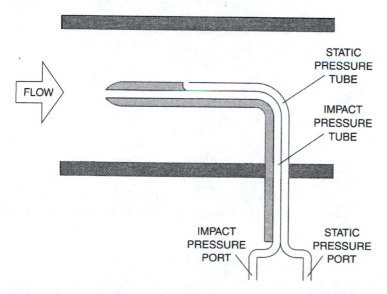

FIGURE 9–23

Pitot Tube Flow Sensor

9-6 Inclination

Inclination sensors are generally called **inclinometers** or **tilt gages.** Inclinometers usually have an electrical output while tilt gages have a visual output (usually a meter or fluid bubble indication). Inclinometers normally have an analog output, but if they have a discrete output they are called **tilt switches.**

The most popular inclinometer is the electrolytic inclinometer. This device consists of a glass tube with three electrodes—one mounted in each end and one in the center. The tube is filled with a non-conductive liquid (such as glycol), which acts as the electrolyte. Since the tube is not completely filled with liquid, there will be a bubble in the tube much like a carpenter's bubble level. As the tube is tilted, the bubble will travel away from the center line of the tube. A typical inclinometer tube is shown in Figure 9–24a. Since there are electrodes in each end of the tube, there are two capacitors formed within the tube—one capacitor is from one end electrode and the center electrode, and the other capacitor is from the other end electrode and the center electrode. As long as the tube is level, the bubble is centered and each capacitor has the same amount of electrolyte between its electrode pair, thus making the two capacitor values equal. However, when the tube is tilted, the bubble shifts position inside the tube. This changes the amount of dielectric (electrolyte) between the electrodes, which causes a corresponding shift in the capacitance ratio between the two capacitors. By connecting the two capacitors in the tube into an ac Wheatstone bridge and measuring the output voltage, the shift in capacitance ratio (and the amount of tilt) can be measured as a change in voltage, and the direction of tilt can be measured by analyzing the direction of phase shift.

When installing the inclinometer, it is extremely important to make sure that the measurement axis of the inclinometer is aligned with the direction of the inclination to be measured. This is because inclinometers are designed to ignore tilt in any direction except the measurement axis (this is called cross-axis rejection or off-axis rejection). The inclinometer must also be mechanically zeroed after installation. This is done using adjustment screws or adjustment nuts as shown in Figure 9–24b.

The inclinometer system output voltage after signal conditioning and calibration is usually graduated in volts/degree or, for the more sensitive inclinometers, volts/arcminute. The signal conditioning and calibration allow the designer to connect the output of the inclinometer system directly to the analog input of a PLC.

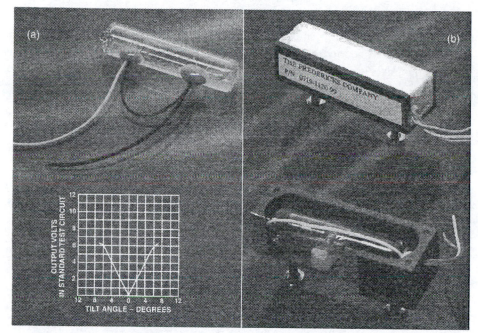

FIGURE 9–24

Inclinometer Tube (a) and Inclinometer Assemblies (b) (Courtesy of the Fredericks Company)

9-7 Acceleration

Acceleration sensors (called **accelerometers**) are used in a variety of applications including aircraft g-force sensors, automotive air bag controls, vibration sensors, and instrumentation for many test and measurement applications. Acceleration measurement is very similar to inclination measurement with the output being graduated in volts/g instead of volts/degree. However, the glass tube electrolytic inclination method described previously cannot be used because the accelerometer must be capable of accurate measurements in any physical position. For example, if we were to orient an accelerometer so that it is standing on end, it should output an acceleration value of 1 g. Glass tube inclinometers will reach a saturation point under extreme inclinations, accelerometers will not.

Acceleration measurement sounds somewhat complicated; however, it is relatively simple. One method to measure acceleration is to use a known mass connected to a pressure strain gage as shown previously in Figure 9–15. Instead of the diaphragm being distorted by fluid or gas pressure, it is distorted by the force from a known mass resting against the diaphragm. When a pressure strain gage is used to measure weight (or acceleration), the device is called a **load cell.**

Figure 9–25 shows three strain gage accelerometers. Figure 9–25a is a one-axis accelerometer, Figure 9–25b shows a two-axis accelerometer, and Figure 9–25c shows a three-axis accelerometer. Note in Figures 9–25b and 9–25c that each measurement axis is marked on the device.

FIGURE 9–25

Strain Gage Accelerometers (Courtesy of Entran Devices, Inc.)

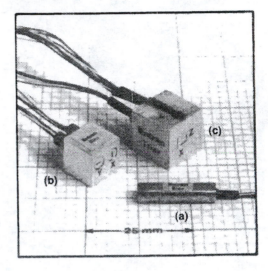

Each accelerometer requires a strain gage bridge compensation resistor and amplifier as shown in Figure 9–26. In this figure, the IMV Voltage Module is an instrumentation amplifier with voltage output. In some applications, the strain gage bridge compensation resistor and amplifier are included in the strain gage module.

FIGURE 9–26

Accelerometer Strain Gage Signal Conditioning (Courtesy of Entran Devices, Inc.)

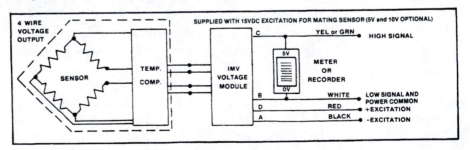

9-8 Angle Position Sensors

Slotted Disk and Opto-Interrupter

When designing or modifying rotating machines, it is occasionally necessary to know when the machine is in a particular angular position, the rotating speed of the machine, or how many revolutions the machine has taken. Although this can be done with an optical encoder, one relatively inexpensive method to accomplish this is to use a simple slotted disk and opto-interrupter. This device is constructed of a circular disk (usually metal) mounted on the machine shaft as shown in Figure 9–27. A small radial slot is cut in the disk so that light from an emitter will pass through the slot to a photo-transistor when the disk is in a particular angular position. As the disk is rotated, the photo-transistor outputs one pulse per revolution. Generally, the slotted disk is painted flat black or is black anodized to keep light scattering and reflections to a minimum.

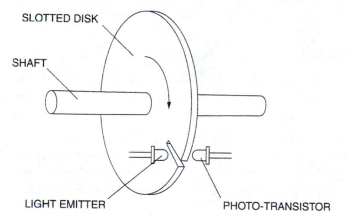

SLOTTED DISK

SHAFT

LIGHT EMITTER   PHOTO-TRANSISTOR

FIGURE 9–27

Slotted Disk and Opto-Interrupter

The slotted disk system can be used to initialize the angular position of a machine. The process of initializing a machine position is called **homing,** and the resulting initialized position is called the **home position.** To do this, the PLC simply turns on the machine's motor at a slow speed and waits until it receives a signal from the photo-transistor. Generally, homing is also a timed operation. That is, when the PLC begins homing, it starts an internal timer. If the timer times out before the PLC receives a home signal, then there is evidently something wrong with the machine (i.e., either the machine is not rotating or the slotted disk system has malfunctioned). In this case, the PLC will shut down the machine and produce an alarm (flashing light, beeper, etc.)

If the slotted disk is used to measure rotating speed, there are two standard methods to do this.

   1. In the first method, the PLC starts a retentive timer when it receives a pulse from the photo-transistor. It then stops the timer on the next pulse. The rotating speed of the machine in RPM is then $S_{rpm} = 60/T$, where T is the time value in the timer after the second pulse. This method works well when the machine is rotating very slowly (<<1 revolution per second), and it is important to have a speed update on the completion of every revolution.

2. In the second method, we start a long timer (say 10 seconds) and use it to enable a counter that counts pulses from the photo-transistor. When the timer times out, the counter will contain the number of revolutions for that time period. We can then calculate the speed, which is $S_{rpm} = C \cdot 60/T$, where C is the value in the counter and T is the preset (in seconds) for the timer used to enable the counter.

Most modern PLCs have built-in programming functions that will do totalizing and frequency measurements. These functions relieve the programmer of writing the math algorithms to perform these measurements for a slotted disk system.

Incremental Encoder

Although the slotted disk encoder works well for once-per-revolution indexing, it may be necessary to measure and rotate a shaft to a precise angular position smaller than 360°. When this is required, an incremental optical encoder may be used.

Consider the slotted disk mechanism described previously, but with the disk replaced by the one shown in Figure 9–28. This is a clear glass disk with black regions etched into the glass surface every 10°.[1] Since the lines are typically small and close together, it is difficult to position a light emitter and photo-transistor

FIGURE 9–28

10° Incremental Optical Encoder Disk and Shadow Mask

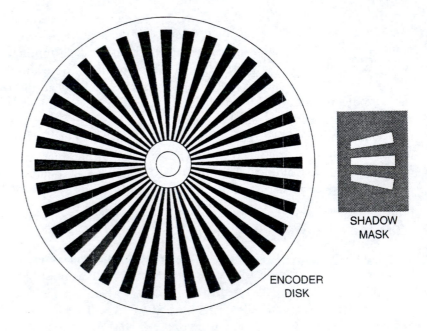

SHADOW MASK

ENCODER DISK

---

[1] The 10-degree increment in this figure is for illustrative purposes only. Actual incremental encoders have *many* more etched lines per revolution.

so that they can detect one line. Therefore, a **shadow mask** is used, which is an opaque material into which slots are cut that match the lines on the disk. When the shadow mask is positioned over the disk, light will pass whenever the clear spaces on the disk are aligned with the slots in the shadow mask, and no light will pass when the black lines are aligned with the shadow mask slots.

For this type of encoder, there are two shadow masks labeled A and B, each with a light emitter on one side of the disk and a corresponding photo-transistor on the opposite side. The shadow mask assemblies are positioned so that when A totally blocks the light, B is ½ of a line width (¼ of a cycle) from blocking the light. Our example encoder has 36 lines, each at a 10° increment. Therefore, as the disk is rotated 10°, each shadow mask will go through one complete dark-to-light-to-dark cycle, as shown in Figure 9–29. Notice in this figure that as the disk is rotated

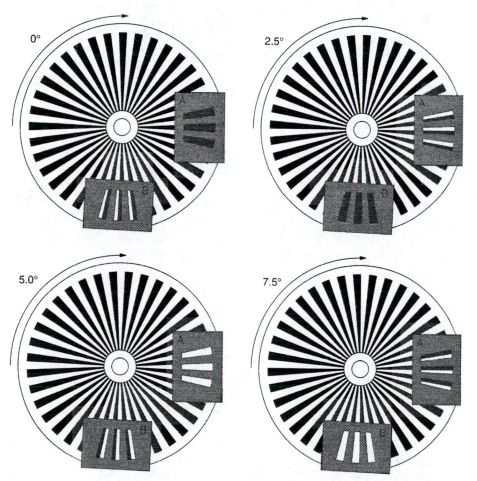

FIGURE 9–29

Incremental Encoder with Shadow Masks A and B, Showing Various Angles of Rotation

clockwise, the light pattern from shadow mask B is the same as it was at shadow mask A 2.5° earlier. Therefore, the signal from the photo-transistor of B lags that of A by 2.5° of encoder shaft rotation. Since the electrical signal from the shadow mask assemblies go through one complete cycle every 10°, the 2.5° degree lag of B corresponds to ¼ of a cycle, or 90 electrical degrees.

If the disk is rotated at a constant angular velocity, both of the photo-transistors will produce a square wave signal (assuming their outputs are conditioned so that they have binary values). However, as the disk is rotated clockwise, the square wave produced by the phase B photo-transistor will appear to lag that of the phase A photo-transistor by approximately 90 electrical degrees. Similarly, if the disk is rotated counterclockwise, the phase B square wave will *lead* phase A by approximately 90°. These relationships are shown in Figure 9–30.

FIGURE 9–30
Incremental Encoder Output Waveforms

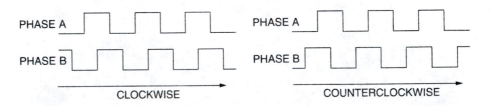

Incremental encoders are specified by the number of pulses per revolution that are produced by either the phase A or phase B output. By dividing the number of pulses per revolution into 360°, we get the number of degrees per pulse (called the **resolution**). This is the smallest change in shaft angle that can be detected by the encoder. For example, a 3,600 pulse incremental encoder has a resolution of 360°/3600 = 0.1°. Incremental encoders are commonly available with up to 72,000 pulses per revolution (0.005° resolution). Even higher resolutions are available by adding mechanical gear assemblies to the encoder shaft.

An incremental encoder can be used to extract three pieces of information about a rotating shaft:

1. By counting the number of pulses received and multiplying the count by the encoder's resolution, the *angular rotation angle* of the shaft can be determined.

2. By viewing the phase relationship between the phase A and phase B outputs, the *direction* of shaft rotation can be found.

3. By counting the number of pulses received from either output during a fixed time period, the *angular velocity* of the shaft (in either radians per second or revolutions per second) can be found.

When an incremental encoder is switched on, it simply outputs a 1 or 0 on its phase A and phase B output lines. This does not provide any initial information about the angular position of the encoder shaft. In other words, the incremental encoder gives **relative position** information, with the reference position being the angle of the shaft when the encoder was energized. The only way an incremental encoder can be used to provide **absolute position** information is for the encoder shaft to be homed after power is applied. This requires some other external device (such as a slotted disk) to provide this home position reference. Some incremental encoders have a third output signal named **home,** which provides one pulse per revolution and can be used for homing the encoder.

Some incremental optical encoders are designed so that the phase A output will lead the phase B output when the encoder shaft is turned clockwise (as with our example) when viewed from the shaft end. However, others operate in just the opposite way. Therefore, designers should consult the technical data for the particular encoder being used. Keep in mind that, should the phase relationship between the phase A and phase B outputs be wrong, it is easily fixed by either swapping the two-phase connections, or by logically inverting one of the two signals.

Example 9–5
Problem:

A 2,880 pulse-per-revolution incremental encoder is connected to a shaft. Its phase A outputs 934 pulses when the shaft is moved to a new position. What is the change in angle in degrees?

Solution:

The resolution of the encoder is $360°/2880 = 0.125°$ per pulse. Therefore the angle change is $0.125 \times 934 = 116.75°$.

Example 9–6
Problem:

A 1,440 pulse-per-revolution incremental encoder outputs a 1,152 Hz square wave from phase A. How fast is the encoder shaft turning in RPMs?

Solution:

First, calculate the rotating speed in revolutions per second. Since the encoder outputs 1,152 pulses per second, it is rotating at $1152/1440 = 0.8$ revolution per second. Now multiply this by 60 seconds to get the rotating speed in RPMs, which is $0.8 \times 60 = 48$ RPMs.

Absolute Encoder

Unlike the incremental encoder, the **absolute encoder** provides digital *values* as an output signal. The output is in the form of a binary word that is proportional to the angle of the shaft. The absolute encoder does not need to be homed because when it is energized, it simply outputs the shaft angle as a digital value.

The absolute encoder is constructed similar to the incremental encoder in that it has an etched glass circular disk with opto-emitters and photo-transistors to detect the clear and opaque areas in the disk. However, the disk has a different pattern etched into it, as shown in Figure 9–31 (note that this is for illustrative purposes only—a 4-bit encoder is of little practical use). The pattern is a simple binary count pattern that has been curved into a circular shape. For the disk shown, there are four distinct rings of patterns, each identified with a numerical weight that is a power of two.

FIGURE 9–31

4-Bit Binary Optical Absolute Encoder Disk

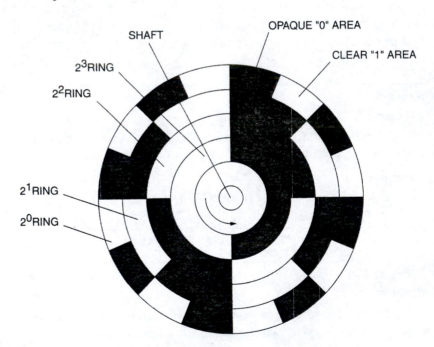

The encoder is constructed so that there is one photo-transistor aligned with each ring on the glass disk. As the shaft and disk are rotated, the photo-transistors output the binary pattern that is etched into the disk. For the disk shown, assuming the shaft is rotated counterclockwise, the output signals would appear as shown in Figure 9–32. Since the encoder disk layout is for 4-bit binary, one revolution of the disk causes an output of sixteen different binary patterns. This means that each of the sixteen patterns will cover an angular range of $360°/16 = 22.5°$. This range of coverage for each output pattern is called the angular resolution. The absolute angle of the encoder shaft can be found by multiplying the binary output of the encoder times the resolution. For example, assume our 4-bit encoder has an output of 1101_2 (decimal 13). The encoder shaft would therefore be at an angle of $13 \times 22.5° = 292.5°$. Because of the relatively poor resolution of this encoder, the shaft could be at some angle between $292.5°$ and $292.5 + 22.5°$.

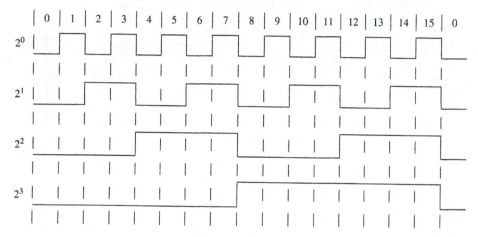

FIGURE 9–32
4-Bit Binary Encoder Output Signals

Example 9–7
Problem:

A 12-bit binary absolute encoder is outputting the number 101100010111. (a) What is the resolution of the encoder, and (b) what is the range of angles indicated by its output?

Solution:

a. A 12-bit encoder will output 2^{12} binary numbers for one revolution. Therefore, the resolution is $360°/2^{12} = 0.087891°$.

b. Converting the binary number to decimal, we have $101100010111_2 = 2839_{10}$. The indicated angle is between $2839 \times 0.087891° = 249.52°$ and $249.52° + 0.087891° = 249.60°$.

One inherent problem that is encountered with binary output absolute encoders occurs when the output of the encoder changes its value. Consider our 4-bit binary encoder when it changes from 7 (binary 0111) to 8 (binary 1000). Notice that in this case, the state of all four of its output bits change value. If we were to capture the output of the encoder while these four outputs are changing state, it is likely that we will read an erroneous value. The reason for this is that because of the variations in slew rates of the photo-transistors and any small alignment errors in the relative positions of the photo-transistors, it is unlikely that all four of the outputs will change at exactly the same instant. For this reason, all binary output encoders include one additional output line called **data valid** (also called **data available,** or **strobe**). This is an output that, as the encoder is rotated, goes false for the very short instant while the outputs are changing state. As soon as the outputs are settled, the data valid line goes true, indicating that it is safe to read the data. This is illustrated in the timing diagram in Figure 9–33.

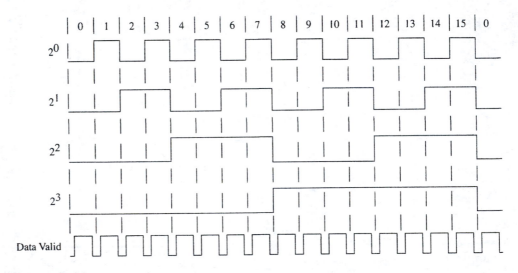

FIGURE 9–33
4-Bit Binary Encoder Output Signals with Data Valid Signal

It is possible to purchase an absolute encoder that does not need a data valid output signal (nor does it have one). In some applications, this is more convenient because it requires one less signal line from the encoder to the PLC (or computer), and the PLC (or computer) can read the encoder output at any time without error. This is done by using a special output coding method that is called **gray code.** Gray code requires the same number of bits to achieve the same resolution as a binary encoder equivalent; however, the counting pattern is established so that, as the angle increases or decreases, no more than one output bit changes at a given time. Many present-day PLCs include math functions to convert gray code to binary, decimal, octal, or hexadecimal.

Like binary code, gray code starts with 000 . . . 000 as the number "zero" and 000 . . . 001 as the number "one." However, from this point, gray code takes a different direction and is unlike binary. Converting from gray code to binary and vice versa is relatively easy by following a few simple steps.

Converting Binary to Gray

1. Write the binary number to be converted and add a leading zero (on the left side).

2. Exclusive-OR each pair of bits in the binary number together, and write the resulting bits below the original number.

Example 9–8
Problem:

Convert 1001110010_2 to gray code.

Solution:

Step 1– 01001110010_2 (add a leading zero)

Step 2– exclusive-OR adjacent bits

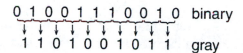

0 1 0 0 1 1 1 0 0 1 0 binary

1 1 0 1 0 0 1 0 1 1 gray

Converting Gray to Binary

1. Write the gray code number to be converted and add a leading zero (on the left side).

2. Beginning with the leftmost digit (the added zero), perform a chain addition of all the bits and write the "running sum" as you go (discard all carrys).

Example 9–9
Problem:

Convert 1101001011 gray to binary.

Solution:

Step 1– 01101001011_G (add a leading zero)

Step 2– 0

$+1 = 1$
$+1 = 0$
$+0 = 0$
$+1 = 1$
$+0 = 1$
$+0 = 1$
$+1 = 0$
$+0 = 0$
$+1 = 1$
$+1 = 0$

The equivalent binary number is 1001110010_2.

It is extremely important to remember that whenever converting gray to binary, or binary to gray, the total number of bits before and after the conversions must be the same. For example, a 16-bit gray code number will *always* convert to a 16-bit binary number and vice versa.

It may seem that gray code would have the same inherent problem with read errors as binary code if data is read while the encoder output is transitioning. However, remember that in gray code, any two adjacent values differ by only one bit change. This means that even if a PLC were to read the data while it is changing, the only possible error will be in the single transitioning bit. Therefore, the only possibility is that the PLC will read the number as one of the two adjacent numbers, hardly a gross error. Therefore, it is unnecessary to strobe the output of a gray code absolute encoder.

Example 9–10
Problem:

A 10-bit gray code optical encoder is outputting the number 0011100101. What is the indicated angle?

Solution:

First, find the resolution of a 10-bit encoder, which is $360°/2^{10} = 0.35156°$. Then convert 0011100101_G to binary using the method described previously. This will result in the binary number 0010111001. Now convert this number to its decimal equivalent, which is 185, and multiply it by the resolution to get $185 \times 0.35156° = 65.04°$.

9-9 Linear Displacement

Potentiometer

A slide potentiometer is the simplest of all linear displacement sensors. Its principle of operation is simply that we apply a voltage to a resistor and then move a slider across the resistor. The voltage appearing on the slider is proportional to the physical position of the slider on the resistor. Generally, in most displacement sensing linear resistors, the resistive element is made from small wire. This makes the unit more durable and less sensitive to variations in temperature and humidity. These are called **slidewire potentiometers.** The most common applications for slidewire potentiometers are in XY plotters and chart recorders.

Linear Variable Differential Transformer (LVDT)

The linear variable differential transformer, or LVDT, is a mature technology that is still very popular. The device operates on the principle that the amount of coupling between the primary and secondary of a transformer depends on the placement of the transformer core material. Consider the LVDT shown in Figure 9–34. Identical coils (equal numbers of turns and equal size wire) L1 and L2 make up the primary of the transformer, with L3 being the secondary. All three coils are wound on a non-metallic tube. L1 and L2 are connected such that the same alternating current passes through both coils but in opposite directions. This means that the magnetic fields produced by L1 and L2 will be equal but opposite. At the exact center between L1 and L2, the magnetic fields from the two coils will cancel and produce a net field of zero. Therefore, the induced voltage in coil L3 will also be zero.

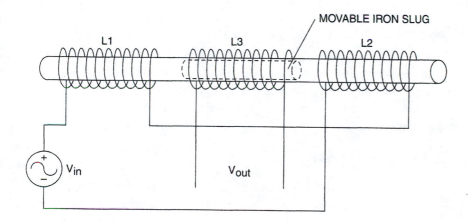

FIGURE 9–34

Linear Variable Differential Transformer (LVDT)

A high permeability iron slug is positioned inside the tube, and a rod from the slug connects to the moving mechanical part to be sensed. As long as the slug remains centered, the flux coupled from coils L1 and L2 into L3 will be equal and opposite and will produce no induced voltage in L3. However, if the slug is moved slightly to the right, the permeability of the slug will lower the reluctance for coil L2 and increase the magnetic flux coupled from L2 to L3. At the same time, less flux from L1 will be coupled to L3. This will cause a small voltage to be induced in L3 that is in-phase with the voltage applied to L2. Moving the slug farther to the right will cause a proportional increase in the amplitude of the voltage induced in L3.

If we move the slug to the left of center, more of the magnetic flux from L1 will be coupled to L3 causing an induced voltage that is in-phase with the voltage applied to L1. Notice that since L1 and L3 are connected in opposite polarity, moving the slug to the left will cause a phase reversal in the voltage induced in L3 as compared to moving the slug to the right. Therefore, by measuring the amplitude of the induced voltage in L3, we can determine the magnitude of the displacement of the slug from the center, and by measuring the phase relationship between the induced voltage in L3 and the applied voltage V_{in}, we can determine whether the direction of displacement is right or left.

Naturally, both the ac amplitude and phase of the voltage V_{out} must be conditioned in order to provide a dc output voltage that is proportional to the displacement of the iron slug. Because of this, modern LVDTs come with built-in signal conditioning electronic circuitry. They are generally powered by dual 15 volt power supplies and produce an output voltage of either -5 to $+5$ volts or -10 to $+10$ volts over the range of mechanical travel.

Ultrasonic

The ultrasonic distance sensor works in the same manner as the ultrasonic proximity sensor. However, instead of a discrete output, the sensor has an analog output that is proportional to the distance from the sensor to the target object. The ultrasonic distance sensor has the same advantages and disadvantages as the ultrasonic proximity sensor.

Glass Scale Encoders

If we were to unwrap the glass disk of a rotary encoder and make it a straight, narrow glass scale, we could easily have an encoder that would, instead of producing angular position information, produce linear displacement information. Like rotary encoders, glass scale encoders are available in both incremental and absolute versions. Their principles of operation are identical to that of their rotary counterparts. Since they are capable of detecting linear movements as small as 0.0001" or less, they are commonly used in applications that require extreme accuracy and repeatability such as numerically controlled mill and lathe machines.

Magnetostrictive Sensors

The magnetostrictive linear displacement sensor is a relatively new technology that has been perfected for making precise measurements of linear motion and position. The system utilizes two fundamental physical properties: (a) whenever a current is passed through a conductor resting in a magnetic field there will be a mechanical force produced, and (b) sound waves travel through a solid material at a predictable velocity.

Consider the cutaway drawing of a magnetostrictive position sensor shown in Figure 9–35. The sensor unit consists of a sensor element head, waveguide, and sensor element protective tube. The sensor element head contains the electronic circuitry to operate the system. The waveguide is fundamentally a length of steel wire. For this application, it must be both conductive and elastic (much like a piano string). The sensor element tube is conductive and provides a protective shell for the waveguide. External to (and separate from) these elements is a toroidal permanent magnet. The magnet is generally attached to the mechanism that moves, while the sensor element head, waveguide, and tube remain stationary. The waveguide tube is inserted through the hole in the toroidal magnet.

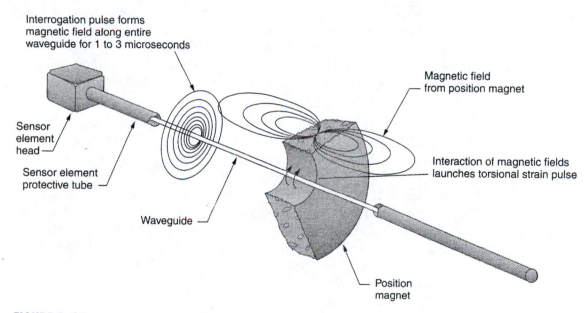

FIGURE 9–35
Magnetostrictive Position Sensor, Cutaway View
(Courtesy of MTS Sensors Division)

In operation, the sensor element head generates an extremely short current pulse that is applied to the waveguide. This pulse will cause a magnetic field to be induced around the waveguide. This magnetic field will interact with the stationary magnetic field produced by the toroidal magnet and cause a very short mechanical pulse to be produced on the waveguide. This is much like plucking the string of a guitar; however, in this case, the mechanical force is a torsional (twisting) force. The mechanical pulse causes a torsional wave to begin traveling down the waveguide toward the sensor element head. The system works much like sonar, except that we have a transmitted electrical signal that produces a mechanical echo.

The sensor element head contains a mechanical transducer that measures torsional motion of the waveguide and converts it to an electrical pulse. Then the electronic circuitry in the sensor element head measures the elapsed time between the current pulse and the returning mechanical pulse. This time is directly proportional to the distance the permanent magnet is located from the sensor element head. The pulse delay is then converted to an analog voltage that appears on the output terminals of the sensor.

The electronic circuitry is generally calibrated to have an output range of 0 to 10 volts. Zero volts corresponds to a distance of zero (i.e., the magnet is located at the sensor head end of the tube), and 10 volts corresponds to the full length of the tube. Sensors are available with various length tubes. As an example, assume we are using a 24" sensor. In this case, 10 volts = 24", and all positions from 0" to 24" are scaled proportionally.

As an application example, consider the arrangement shown in Figure 9–36. This shows a cutaway view of a hydraulic cylinder on the right and a magnetostrictive sensor on the left. The shaft of the hydraulic cylinder has been bored to make room for the sensor's waveguide tube. The toroidal permanent magnet is fastened to the face of the piston inside the cylinder. As the hydraulic cylinder extends or contracts, the magnet will move to the right and left on the waveguide tube. This will cause the sensor to output a voltage that is proportional to the mechanical extension of the hydraulic cylinder. This method is widely used in flight simulators to precisely and smoothly control the hydraulic cylinders that move the platform.

Example 9–11
Problem:

A 36" magnetostrictive sensor has a specified output voltage range of 0 to 10 volts. If it is outputting 8.6 volts, how far is the magnet positioned from the sensor head?

Solution:

The answer can be found by using a simple ratio: 10 V is to 36" as 8.6 V is to X". Therefore, X = 30.96".

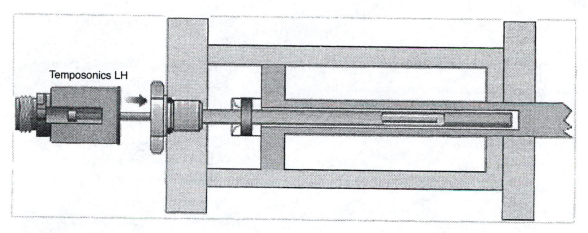

FIGURE 9–36
Magnetostrictive Sensor Measurement of Hydraulic Cylinder Displacement
(Courtesy of MTS Sensors Division)

Summary

The encoders, transducers, and sensors we have explored in this chapter are the most popular and most commonly used in control systems. However, as with all sensor technology, the types of available off-the-shelf devices continuously evolve and the selections expand. Therefore, it is important for the control systems designer to stay abreast of the currently available equipment by maintaining an up-to-date library of manufacturers' catalogs and to browse the catalogs obtained to gain an awareness of these new technologies.

Review Questions

1. Explain the difference between a transducer and a sensor.

2. When a bi-metallic strip is heated, it bends in the direction of the side that has the _____ (high or low) coefficient of expansion.

3. As applied to discrete sensors, what is meant by the term hysteresis?

4. What is the Seebeck voltage?

5. What metals make up the alloys a) chromel, b) alumel, and c) constantan?

6. Explain why a thermocouple cannot be connected using standard copper hookup wire.

7. Why is platinum the most popular metal used to make RTDs?

8. A 200 ohm platinum RTD measures 222 ohms. What is its temperature in Celsius?

9. What would be the resistance of a 100 ohm platinum RTD that is immersed in boiling water (100° C)?

10. An integrated circuit temperature probe is rated at 0–10 volts dc over a 0–175° C range. What is the temperature when the output voltage is 3.65 volts dc?

11. A N/O float switch closes when the liquid level is _____ (low or high).

12. A 100 mm long metal rod is compressed and measures 97 mm. What is the strain?

13. A 20' support I-beam is loaded in compression lengthwise so that the microstrain is $\mu\varepsilon = 225$. What is the length of the loaded beam?

14. A 1k ohm strain gage undergoes a resistance change of 1.575 ohms when it is subjected to a microstrain of $\mu\varepsilon = 450$. What is the gage factor of the strain gage?

15. A 1k ohm strain gage is connected into a bridge circuit with the other three resistors $R_1 = R_2 = R_3 = 1$k ohms. The bridge is powered by a 10 volt dc power supply. The strain gage has a gage factor $k = 1.5$. The bridge is balanced ($V_{out} = 0$ volts) when the strain $= 0$. What is the strain when $V_{out} = +200$ microvolts?

16. A 0 to +500 psi pressure transducer has a calibration factor of 100 psi/volt.
(a) What is the pressure if the transducer output is 1.96 volts? (b) What is the full scale output voltage of the transducer?

17. A 0 to 1,500 psi pressure transducer has a full scale output voltage of 5 volts.
(a) What will be the output voltage when it is subjected to a pressure of 975 psi?

(b) What is the transducer's calibration factor?

18. Water is flowing at 10 mph in a square concrete drainage canal that is 10' wide. The water in the canal is 5' deep. What is the flow in ft³/sec?

19. What is an advantage in using a thermal dispersion flow switch as opposed to other types of flow switches?

20. You have installed into a water pipe a turbine flow sensor that has a calibration factor of 60 pulses/gallon. When water is flowing in the pipe the sensor outputs 470 pulses/minute. What is the flow rate in gallons/minute?

21. A turbine flow sensor having a calibration factor of 2 pulses/gallon is connected to the discrete input of a PLC. The PLC is programmed to measure the time between pulses (the period). If it measures 1.75 seconds per pulse, what is the flow rate in gallons/minute? Show the equation that the PLC would need to convert the period T into the flow Q in gallons/minute.

22. A slotted disk and opto-interrupter as shown in Figure 9–27 outputs pulses at 62 Hz. What is the rotating speed of the disk in rpm?

23. A slotted disk and opto-interrupter as shown in Figure 9–27 outputs one pulse every 3.75 seconds. What is the rotating speed of the disk in rpm?

24. A 3,600 pulse incremental encoder outputs a 2.5 kHz square wave. What is (a) the resolution of the encoder, and (b) the speed of rotation?

25. A 10-bit absolute optical encoder outputs the octal number 137_8. What is its shaft position with respect to mechanical zero degrees?

26. A 10-bit absolute optical encoder is outputting the gray code number 1000110111_G. What is its shaft position with respect to mechanical zero?

27. A 16-inch magnetostrictive sensor has a full scale output of 10 volts. When it is outputting 3.55 volts, how far is the magnet from the sensor element head?

28. A 24-inch magnetostrictive sensor has a full scale output of 10 volts. If the magnet is 13.75 inches from the sensor element head, what will be the output voltage?

Closed-Loop and PID Control

Objectives

Upon completion of this chapter, you will know:

- the basic parts of a simple closed-loop control system.
- why proportional control alone is usually inadequate to maintain a stable system.
- the effect that the addition of derivative control has on a closed-loop system.
- the effect that the addition of integral control has on a closed-loop system.
- how to tune a PID control system.

Introduction

One of the greatest strengths in using a programmable machine control, such as a PLC, is in its capability to adapt to changing conditions. When properly designed and programmed, a machine control system is able to sense that a machine is not operating at the desired or optimum conditions and can automatically make adjustments to the machine's operating parameters so that the desired performance is maintained, even when the surrounding conditions are less than ideal. In this chapter, we will discuss various methods of controlling a closed-loop system and the advantages and disadvantages of each.

10–1 Simple Closed-Loop Systems

When a control system is designed such that it receives operating information from the machine and makes adjustments to the machine based on this operating information, the system is said to be a **closed-loop sys-**

tem, as shown in Figure 10–1. The operating information that the controller receives from the machine is called the **process variable (PV)** or **feedback,** and the input from the operator that tells the controller the desired operating point is called the **setpoint (SP).** When operating, the controller determines whether the machine needs adjustment by comparing (by subtraction) the setpoint and the process variable to produce a difference (the difference is called the **error**). The error is amplified by a **proportional gain**[1] factor k_p in the **proportional gain amplifier** (sometimes called the **error amplifier**). The output of the proportional gain amplifier is the **control variable (CV),** which is connected to the controlling input of the machine. The controller takes appropriate action to modify the machine's operating point until the control variable and the setpoint are nearly equal.

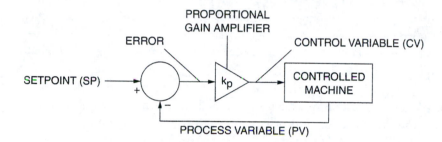

FIGURE 10–1

Simple Closed-Loop Control System

It is important to recognize that some closed-loop systems do not need to be completely proportional (or analog). They can be partially discrete. For example, the thermostat that controls the heating system in a home is a discrete output device; that is, it provides an output that either switches the heater fully on or completely off. The setpoint for the system is the temperature setting that the homeowner can adjust, and the process variable is the room temperature. If the PV is lower than the SP, the thermostat switches the CV to the ON state, in this case a discrete ON signal that switches on the heater. The system adapts to external conditions; that is, on warm days when the house is comfortable, the thermostat keeps the heater off, and on very cold days, the thermostat operates the heater more often and for longer periods of time. The result is that, despite the changing outdoor temperature, the indoor temperature remains relatively constant.

Some closed-loop control systems are totally proportional. Consider, for example, the automobile cruise control. The operator "programs" the system by setting the desired vehicle speed (the SP). The controller then compares this value to the

---

[1] In some control systems the term **proportional band** (with variable P) is used instead of proportional gain. The proportional band is a percentage of the inverse of the proportional gain, or $P = 100 / k_p$.

actual speed of the vehicle (the PV) and produces a CV. In this case, the CV results in the accelerator pedal being adjusted so that the vehicle speed is either increased or decreased as needed to maintain a nearly constant speed near the SP, even if the auto is climbing or descending hills. The CV signal that controls the accelerator pedal is not discrete, nor would we want it to be. In this application, having a discrete CV signal would result in some very abrupt speed corrections and an uncomfortable ride for the passengers.

When a digital control device (such as a PLC) is used in a control system, the closed-loop system may be partially or totally digital. In this case, it still functions as a proportional system, but instead of the signals being voltages or currents, they are digital bytes or words. The error signal is simply the result of digitally subtracting the SP value from the PV value, which is then multiplied by the proportional gain constant k_p. Although the end result can be the same, there are some inherent advantages in using a totally digital system. First, since all numerical processing is done digitally by a microprocessor, the calibration of the fully digital control system will never drift with temperature or over time. Second, since a microprocessor is present, it is relatively easy to have it perform more sophisticated mathematical functions on the signals such as digital filtering (called **digital signal processing, discrete signal processing,** or **DSP**), averaging, numerical integration, and numerical differentiation. As we will see in this chapter, performing advanced mathematical functions on the closed-loop signals can vastly improve a system's response, accuracy, and stability. Whenever the closed-loop control is performed by a PLC, the actual control calculations are generally performed by a separate co-processor so that the main processor can be freed to solve the ladder program at high speed. Otherwise, adding closed-loop control to a working PLC would drastically slow the PLC scan rate.

10–2 Problems with Simple Closed-Loop Systems

Although the preceding explanation is intended to give the reader an understanding of the fundamentals of closed-loop control systems, unfortunately only a very few types of closed-loop systems will work correctly when designed as shown in Figure 10–1. The reason is that in order for the machine's operating point to be near to the value of the SP, the proportional gain k_p must be high. However, when a high gain is used, the system becomes unstable and will not adjust its CV correctly. Additionally, if the controlled machine has a delay between the time a CV signal is sent to the machine and the time the machine responds, the control system will tend to overcompensate and over-correct for the error.

To see why these are potential problems, consider a closed-loop system that controls the speed of a dc motor as shown in Figure 10–2. In this system, the output of the proportional gain amplifier powers the dc motor. The PV for the system is provided by a tacho-generator connected to the motor shaft. The tacho-generator simply outputs a dc voltage proportional to the rotating speed of the shaft. It appears that if we make the SP the same as the tacho-generator's output (the PV) at the desired speed, the controller will operate the motor at that speed. However, this is not the case.

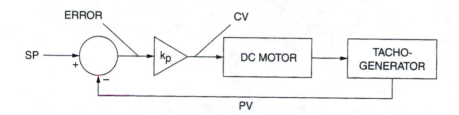

FIGURE 10–2
Simple Closed-Loop dc Motor Speed Control System

When the operator inputs a new SP value to change the motor speed, the control system begins automatically adjusting the motor's speed in an attempt to make the PV match the SP. However, if the proportional gain amplifier has a low gain k_p, the response to the new SP is slow, sluggish, and inaccurate. The reason for this is that as soon as the motor begins accelerating, the tacho-generator begins outputting an increasing voltage as the PV. When this increasing PV voltage is subtracted from the fixed SP, it produces a decreasing error. This means the CV will also decrease, which, in turn, will cause the motor speed to increase at a slower rate. This causes the response to be sluggish. Additionally, in our example, let us assume that in order to operate the motor in the desired direction of rotation the CV must be a positive voltage. This means the error must also be some positive voltage. The only way the error voltage can be positive is for the PV voltage to be less than the SP. In other words, the motor speed will "level off" at some value that is less than the SP. It will never reach the desired speed.

Figure 10–3 is a graph of the speed of a dc motor with respect to time as the motor is accelerated from zero to a SP of 1,000 rpm using a closed-loop control with low proportional gain. Notice how the motor acceleration is reduced as the motor speed increases, which causes the system to take over 90 seconds to settle, and notice that the final motor speed is approximately 350 rpm below the desired setpoint speed of 1,000 rpm. This error is called **offset.**

FIGURE 10–3

Motor Speed Control Response with Low Proportional Gain

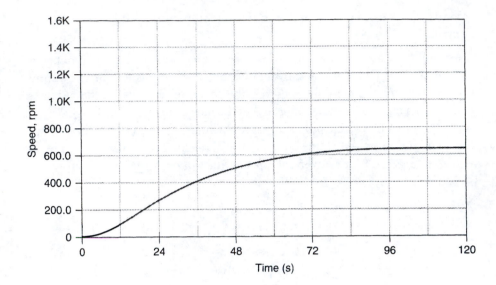

In an attempt to improve both the sluggish response and the offset in our motor speed control, we will now increase the proportional gain k_p. Figure 10–4 is a graph of the same system with the proportional gain k_p doubled. Notice in this case that the motor speed responds faster and the final speed is closer to the SP than that in Figure 10–3. However, this system still takes more than a minute to settle, and the offset is more than 200 rpm below the SP.

FIGURE 10–4

Motor Speed Control Response with Moderate Proportional Gain

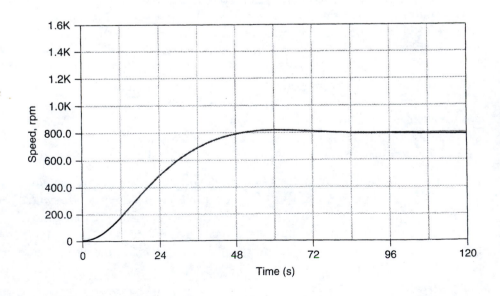

Since doubling the proportional gain k_p seemed to help the response time and the offset of our system, we will now try a large increase in the proportional gain. Figure 10–5 shows the response of our system with the gain k_p increased by a factor of 10. Notice here that the offset is smaller (approximately 25 rpm below the SP); however, the response now oscillates to both sides of the SP before finally settling. This decaying oscillation is called **hunting** and in some systems is generally undesirable. It can potentially damage machines with the overstress of mechanical systems and the overspeed of motors. In addition, it is counterproductive because although our motor speed increased rapidly, the system still required nearly two minutes to settle.

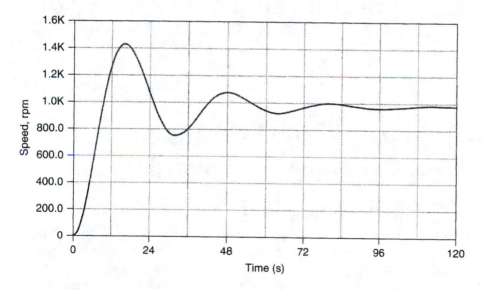

FIGURE 10–5

Motor Speed Control Response with High Proportional Gain

If we increase the proportional gain even more, the system becomes unstable. Figure 10–6 shows this condition, which is called **oscillation**. It is extremely undesirable and, if ignored, will likely be destructive to most closed-loop electro mechanical systems. Any further increase in the proportional gain will cause higher amplitudes of oscillations.

As the previous example illustrates, attempting to improve the performance of a closed-loop system by simply increasing the proportional gain k_p has lackluster results. Some performance improvement can be achieved up to a point, but any further attempts to increase the proportional gain result in an unstable system. Therefore, some other method must be used to "tune" the system to achieve more desirable levels of performance.

FIGURE 10–6

Motor Speed Control Response with Very High Proportional Gain

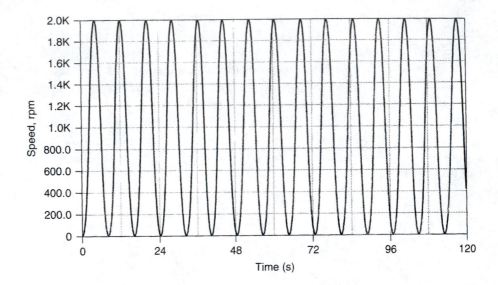

10-3 Closed-Loop Systems Using Proportional, Integral, and Derivative (PID)

In the example previously used, there were three fundamental problems with the simple system using only proportional gain. First, the system had a large offset; second, it was slow to respond to changes in the SP (both of these problems were caused by a low proportional gain k_p); and third, when the proportional gain was increased to reduce the offset and minimize the response time, the system became unstable and oscillated. It was impossible to simultaneously optimize the system for low offset, fast response, and high stability by tuning only the proportional gain k_p.

To improve on this arrangement, we will add two more functions to our closed-loop control system, which are an integral function $k_i \int$ and a derivative function $k_d \frac{d}{dt}$, as shown in Figure 10–7. Notice that in this system the error signal is amplified by k_p and then applied to the integral and derivative functions. The outputs of the proportional gain amplifier, k_p, the integral function, $k_i \int$, and the derivative function, $k_d \frac{d}{dt}$, are added together at the summing junction to produce the CV. The values of k_p, k_i, and k_d are multiplying constants that are adjusted by the system designer and are almost always set to a positive value or zero. If a function is not needed, its particular k value is set to zero.

For the proportional k_p function, the input is simply multiplied by k_p. For the integral k_i function, the input is integrated (by taking the integral) and then multiplied by k_i. For the derivative k_d function, the input is differentiated (by taking the derivative) and then multiplied by k_d. There is some interaction between these three functions; however, generally speaking, each of them serves a specific and unique purpose in our system. Although the names of these functions (integral and derivative) imply the use of calculus, an in-depth knowledge of calculus is not necessary to understand and apply them.

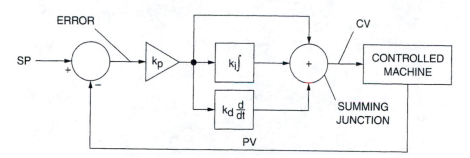

FIGURE 10–7
Closed-Loop Control System with Ideal PID

The PID system shown in Figure 10–7 is called an **ideal PID** and is the most commonly used PID configuration for control systems. There is another popular version of the PID called the **parallel PID** or the **electrical engineering PID** in which the three function blocks (proportional, integral, and derivative) are connected in parallel, as shown in Figure 10–8. When properly tuned, the parallel PID performs identical to the ideal PID. However, the values of k_i and k_d in the parallel PID will be larger by a factor of k_p because the input to these functions is not pre-amplified by k_p as in the ideal PID.

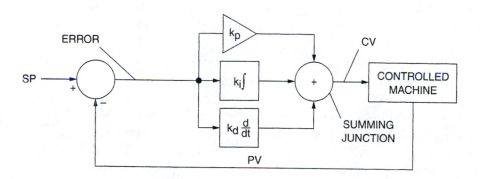

FIGURE 10–8
Closed-Loop Control System with Parallel PID Type

10-4 Derivative Function

By definition, a true derivative function outputs a signal that is equal to the graphical slope of the input signal. For example, as illustrated in Figure 10–9a, if we input a linear ramp waveform (constant slope) into a derivative function, it will output a voltage that is equal to the slope m of the ramp, as shown in Figure 10–9b. For a ramp waveform, we can calculate the derivative by simply performing the "rise divided by the run" or m = $\Delta y/\Delta t$, where Δy is the change in amplitude of the signal during the time period Δt. As Figure 10–9b shows, our ramp has a slope of $+0.5$ during the time period 0 to 2 seconds, and a slope of -0.5 for the time period 2 to 4 seconds.

FIGURE 10–9

Slope (Derivative) of a Ramp Waveform

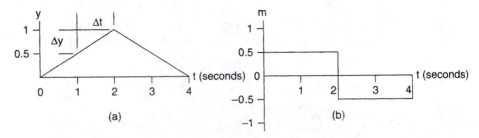

As another example, if we input any constant dc voltage into a derivative function, it will output zero because the slope of all dc voltages is zero. Although this seems simple, it becomes more complicated when the input signal is neither dc nor a linear ramp. For example, if the input waveform is a sine wave, it is difficult to calculate the derivative output because the slope of a sine wave constantly changes with time. Although the exact derivative of a sine wave (or any other waveform) can be determined using calculus, in the case of control systems, calculus is not necessary. This is because it is not necessary to know the exact value of the derivative for most control applications (a close approximation will suffice), and there are alternate ways to approximate the derivative without using calculus. Since the derivative function is generally performed by a digital computer (usually a PLC or a PID co-processor) and the digital system can easily, quickly, and repeatedly calculate $\Delta y/\Delta t$, we may use this sampling approximation of the derivative as a substitute for the exact continuous derivative, even when the error waveform is non-linear. When a derivative is calculated in this fashion, it is usually called a **discrete derivative, numerical derivative,** or **difference function.** Although the function we will be using is not a true derivative, we will still call it a "derivative" for brevity. Even if the derivative function is performed electronically instead of digitally, the function is still not a true derivative because it is usually bandwidth limited (using a low-pass filter) to exclude high frequency components of the signal. The reason is that the slopes of high frequency signals can be extremely steep (high values of m), which will cause the derivative circuitry to output extremely high and erratic voltages. This would make the entire control system overly sensitive to noise and interference.

We will now apply the derivative function to our motor control system as shown in Figure 10–10. For this exercise, we will be adjusting the proportional gain k_p and the derivative gain k_d only. The integral function will be temporarily disabled by setting k_i to zero.

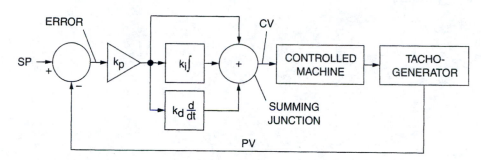

FIGURE 10–10
Closed-Loop Control System with Ideal PID

Previously, when we increased the proportional gain of our simple, closed-loop dc motor speed system example, it began to exhibit instability by hunting. This particular instance was shown in Figure 10–5, and for easy reference, it is shown again in Figure 10–11.

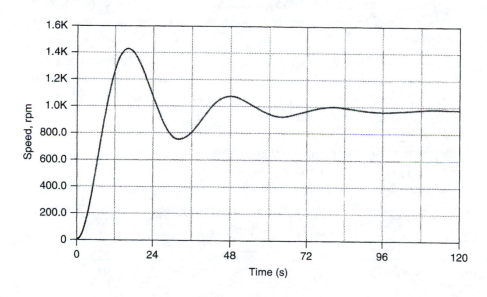

FIGURE 10–11
Motor Speed Control Response with High k_p Only

Without changing the proportional gain k_p, we will now increase the derivative gain k_d a small amount. The resulting response is shown in Figure 10–12.

FIGURE 10–12

dc Motor Speed Control with High k_p and Low Derivative Gain k_d

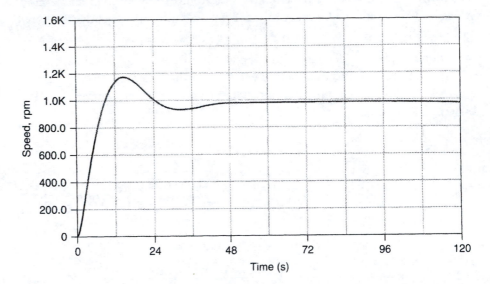

The reader is invited to study and compare Figures 10–11 and 10–12 for a few moments. In particular, notice the way the motor speed accelerates at time $t = 0^+$, the top speed that the motor reaches while the system is hunting, and the length of time it takes the speed to settle.

To see why the addition of derivative gain made such a dramatic improvement in the performance of the system we need to consider the dynamics of the system at certain elapsed times.

First, at time $t = 0$, the SP is switched from zero to a voltage corresponding to a motor speed of 1,000 rpm. Since the motor is not rotating, the tacho-generator output will be zero and the PV will be zero. Therefore, the error voltage at this instant will be identical in amplitude and waveshape to the SP. At time zero, when the SP is switched on, the waveshape will have a very high risetime (i.e., a very high slope), which, in turn, will cause the derivative function to output a very large positive signal. This derivative output will be added to the output of the proportional gain function to form the CV. Mathematically, the CV at time $t = 0^+$ will be

$$CV(t = 0^+) = k_p \, SP + k_d \, m_{SP} \qquad (10\text{–}1)$$

where m_{SP} is the slope of the SP waveform. Since the slope m_{SP} is extremely large at $t = 0^+$, the CV will be large and cause the motor to begin accelerating very rapidly. This difference in performance can be seen in the response curves. In Figure 10–11, the motor hesitates before accelerating, mainly due to starting friction (called **stiction**) and motor inductance, while in Figure 10–12 the motor immediately accelerates due to the added "kick" provided by the derivative function.

Next, we will consider how the derivative function reacts at motor speeds above zero. Since the SP is a constant value and the error is equal to the SP minus the PV, the slope of the error voltage is going to be the opposite polarity of the slope of the PV. In other words, the error voltage will increase when the PV decreases, and the error voltage will decrease when the PV increases. Since the waveshape of the PV is the same as the waveshape of the speed, we can conclude that the slope of the error voltage will be the same as the negative of the slope of the speed curve. Additionally, since the output of the derivative function is equal to the slope of the error voltage, then we can also say that the output of the derivative function will be equal to the negative of the slope of the speed curve (for values of time greater than zero). Mathematically, this can be represented as

$$CV(t > 0) = k_p (SP - PV) + k_d\, m_{(SP - PV)} \qquad \textbf{(10–2)}$$

Since the SP is constant, its slope will be zero. Additionally, as we concluded earlier, the PV is the same as the speed. Therefore, we can simplify our equation to be

$$CV(t > 0) = k_p (SP - speed) - k_d\, m_{speed} \qquad \textbf{(10–3)}$$

In other words, the CV is reduced by a constant k_d times the slope of the speed curve. Therefore, as the motor accelerates, the derivative function reduces the CV and attempts to reduce the rate of acceleration. This phenomenon is what reduced the motor speed overshoot in Figure 10–12 because the control system has "throttled back" on the motor acceleration. In a similar manner, as the motor speed decelerates, the negative speed slope causes the derivative function to increase the CV in an attempt to reduce the deceleration rate.

Since the derivative function tends to dampen the motor's acceleration and deceleration rates, the amount of hunting required to bring the motor to its final operating speed is reduced, and the system settles more quickly. For this reason, the derivative function is sometimes called **rate damping.** If the machine is diesel, gasoline, steam, or gas turbine powered, this is sometimes called **throttle damping.** Also, if the hunting can be reduced by increasing k_d, we can then increase the proportional gain k_p to help further reduce the offset without encountering instability problems.

It would seem logical that if the derivative function can control the rate of acceleration to reduce overshoot and hunting, then we should be able to further improve the motor control performance shown in Figure 10–12 by an additional increase in k_d. Figure 10–13 shows the response of our motor control system where k_d has been increased by a factor of four. Note that now there is only a slight overshoot, and the system settles very quickly. It is possible to achieve even faster response of the system by further increasing the proportional and derivative gain constants k_p and k_d. However, the designer should take caution in doing so because the electrical voltage, and current transients, and the mechanical force transients can become excessive with potentially damaging results.

FIGURE 10–13

DC Motor Speed Control with High k_p and Moderate Derivative Gain k_d

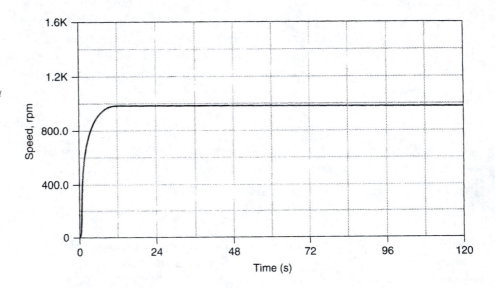

Although the transient response of our motor control system has been vastly improved, the offset is still present. By increasing the derivative constant k_d, we eliminated the overshoot and hunting, but this had no effect on the offset. As indicated in Figure 10–13, the proportional and derivative functions nearly drove the motor speed to the proper value, but then the speed began to drop. As we will investigate next, in order to reduce this offset, the integral function must be used.

Based on the results of our investigations of the derivative function in a PID closed-loop control system, we can make the following general conclusion: *In a properly tuned PID control system, the derivative function improves the transient response of the system by reducing overshoot and hunting. A byproduct of the derivative function is the ability to increase the proportional gain with increased stability, reduced settling time, and reduced offset.*

10-5 Integral Function

The true integral of a function is defined as the graphical area contained in the space bordered by a plot of the function and the horizontal axis. Integrals are cumulative; that is, as time passes, the integral keeps a running sum of the area outlined by the function being integrated. To illustrate this, we will take the integral of the pulse function shown in Figure 10–14a. This function switches from 0 to 2 at time $t = 0$, switches back to 0 at time $t = 1$ second, then at time $t = 2$ seconds switches to -3, and finally back to 0 at $t = 3$ seconds. The integral of this waveform is shown in Figure 10–14b.

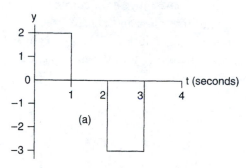
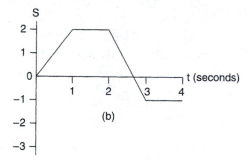

FIGURE 10–14
Accumulated Area (Integral) of a Pulse Waveform

When the input waveform in Figure 10–14a begins at time $t = 0$, we will assume that the integral starting value is zero. As time increases from 0 to 1 second, the area under the waveform increases, which is indicated by the rising waveform in Figure 10–14b. At 1 second when the waveform switches off, the total accumulated area is 2. Since the input is 0 between $t = 1$ s and $t = 2$ s, no additional area is accumulated. Therefore, the accumulated area remains at 2 as indicated by the output waveform in Figure 10–14b between $t = 1$ s and $t = 2$ s. At $t = 2$ s, the input switches to -3, and the integral begins adding negative area to the total. This causes the output waveform to go in the negative direction. Since the area under the negative pulse of the input waveform between $t = 2$ s and $t = 3$ s is -3, the output waveform will decrease by 3 during the same time period.

Notice that the output waveform in Figure 10–14b ends at a value of -1. This is the total accumulated area of the input waveform in Figure 10–14a during the time period $t = 0$ s to $t = 4$ s. In fact, we can determine the total accumulated area at any time by reading the value of the integral in Figure 10–14b at the desired time. For example, the total accumulated area at $t = 0.5$ second is $+1$, and the total accumulated area at $t = 2.5$ seconds is $+0.5$.

For the example waveform in Figure 10–14a, the integral process seems simple. However, as with the derivative, if we wish to take the exact integral of a waveform that is nonlinear, such as a sine wave, the problem becomes more complicated and requires the use of calculus. For a control system (such as a PLC), this would be a heavy mathematical burden. So to lessen the burden, we instead have the PLC sample the input waveform at short, evenly spaced intervals and calculate the area by multiplying the height (the amplitude) by the width (the time interval between samples), and then summing the rectangular slices, as shown in Figure 10–15. Doing so creates an approximation of the integral called the **numerical integral** or **discrete integral.** (It should be noted that in Figure 10–15 the integral of a sine wave over one complete cycle, or any number of complete cycles, is zero because the algebraic sum of the positive slices and negative slices is zero.) As we will see, in a PID control system, the integral function is used to minimize the offset. Since it will reset the system so that the SP and PV are equal, the integral function is usually called the **reset.**

FIGURE 10–15

Discrete Integral of a Sine Wave

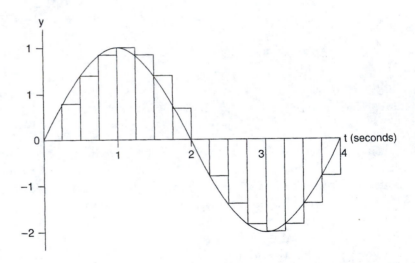

The error incurred in taking a numerical integral when compared to the true (calculus) integral depends upon the sampling rate. A slower sampling rate results in fewer samples and larger error, while a faster sampling rate results in more samples and smaller error.

When used in a PID control system, this type of numerical integrator has an inherent problem. Since the integrator keeps a running sum of area slices, it will retain sums beginning with the time the system is switched on. If the control system is switched on and the machine itself is not running, the constant error can cause extremely high values to accumulate in the integral before the machine is started. This phenomenon is called **reset wind-up** or **integral wind-up.** If allowed to accumulate, these high integral values can cause unpredictable responses at the instant the machine is switched on that can damage the machine and injure personnel. To prevent reset wind-up, when the CV reaches a predetermined limit, the integral function is no longer calculated. This will keep the value of the CV within the upper and lower limits, both of which are specified by the PLC programmer. In some systems, these CV limits are called **saturation,** or **output (CV) min** and **output (CV) max,** or **batch unit high limit** and **batch unit preload.**

In a closed-loop system, whenever there is an offset in the response, there will be a non-zero error signal. This is because the PV and the SP are not equal. Since the integral function input is connected to the same error signal as the derivative function, the integral will begin to sum the error over time. For large offsets, the integral will accumulate rapidly and its output will increase quickly. For smaller offsets, the integral output will change more slowly. However, note that as long as the offset is non-zero, the integral output will be changing in the direction that will reduce the offset. Therefore, in a closed-loop PID control system we can reduce the offset to near zero by increasing the integral gain constant k_i to some positive value.

Therefore, we will now attempt to reduce the offset in our motor speed control example to zero by activating the integral function. Figure 10–16 shows the result of increasing k_i. Note that the transient response has changed little, but after settling, the offset is now zero.

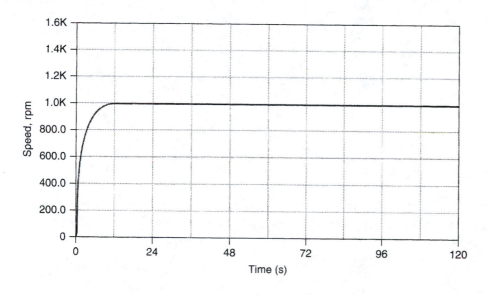

FIGURE 10–16

DC Motor Speed Control with High k_p, Moderate Derivative Gain k_d, and Low Integral Gain k_i

It is not advisable to make the integral gain constant k_i excessively high. Doing so causes the integral and proportion functions to begin working against each other, which will make the system more unstable. Systems with excessive values of k_i will exhibit overshoot and hunting and may oscillate.

To summarize the purpose of the integral function, we can now make the following general statement: *In a closed-loop PID system, the integral function accumulates the system error over time and corrects the error to be zero or nearly zero.*

10-6 The PID in Programmable Logic Controllers

Although the general PID concepts and the effects of adjusting the k_p, k_i, and k_d constants are the same, when using a programmable logic controller to perform a PID control function, there are some minor differences in the way the PID is adjusted. The PID control unit of a PLC performs all the necessary PID calculations on an iterative basis. That is, the PID calculations are not done continuously, but are triggered by a timing function. When the timing trigger occurs, the PV and SP are sampled once and digitized, and then all of the proportional, integral, and derivative functions are calculated and summed to produce the CV. The PID then pauses, waiting for the next trigger.

The exact parallel PID expression is

$$CV = k_p\left[(SP - PV) + k_i\int_0^t (SP - PV)dt + k_d\frac{d}{dt}(SP - PV)\right] \quad (10\text{–}4)$$

However, since the PLC performs discrete integral and derivative calculations based on the sampling time interval Δt, the "discrete" PID performed by a PLC is

$$CV = k_p\left[(SP - PV) + k_i\sum_0^t (SP - PV)\Delta t + k_d\frac{\Delta(SP - PV)}{\Delta t}\right] \quad (10\text{–}5)$$

In the numerical integral portion of Equation 10–5, since $SP - PV$ is the error signal, then $\sum_0^t (SP - PV)\Delta t$ is the sum of the areas (the error times the time interval) as illustrated in Figure 10–15, beginning from the time the system is switched on ($t = 0$) until the present time t. In a similar manner, the numerical derivative $\frac{\Delta(SP - PV)}{\Delta t}$ is the slope of the error signal (the rise divided by the run).

The numerator term $\Delta(SP - PV)$ is the present error signal minus the error signal measured in the most recent sample.

Additionally, in order to make the PID tuning more methodical (as will be shown), most PLC manufacturers have replaced k_i with T_i, which is termed the **reset time constant,** or **integral time constant.** The reset time constant T_i is $1/k_i$, or the inverse of the integral gain constant.[2] For similar reasons, the derivative gain constant k_d has been replaced with the **derivative time constant** T_d, where $T_d = k_d$. Therefore, when tuning a PLC-operated PID controller, the designer will be adjusting the three constants k_p, T_i, and T_d. Using these constants, our mathematical expression becomes

$$CV = k_p \left[(SP - PV) + \frac{1}{T_i} \sum_0^t (SP - PV)\Delta t + T_d \frac{\Delta(SP - PV)}{\Delta t} \right] \quad \textbf{(10–6)}$$

Some controlled systems respond very slowly. For example, consider the case of a massive oven used to cure the paint on freshly painted products. Even when the oven heaters are fully on, the temperature rate of change may be as slow as a fraction of a degree per minute. If the system is responding this slowly, it would be a waste of processor resources to have the PID continually update at millisecond intervals. For this reason, some of the more sophisticated PID controllers allow the designer to adjust the value of the sampling time interval Δt. For slower responding systems, this allows the PID function to be set to update less frequently, which reduces the mathematical processing burden on the system.

10-7 Tuning the PID

Tuning a PID controller system is a subjective process that requires the designer to be very familiar with the characteristics of the system and the desired response of the system to changes in the setpoint input. The designer must also take into account the changes in the response of the system if the load on the machine changes. For example, if we are tuning the PID for a subway speed controller, we can expect that the system will respond differently depending on whether the subway is empty or fully loaded with passengers. Additional PID tuning problems can occur if the system being controlled has a nonlinear response to CV inputs (e.g., using field current control for controlling the speed of a shunt dc

---

[2] If no integral action is desired, T_i is set to zero. Mathematically, this is nonsense because if $T_i = 0$, $k_i = \infty$, which would make the integral gain infinite. However, PID systems are programmed to recognize $T_i = 0$ as a special case and simply switch off the integral calculation.

motor). If PID tuning problems occur because of system non-linearity, one possible solution would be to consider using a fuzzy logic controller instead of a PID. Fuzzy logic controllers can be tuned to match the non-linearity of the system being controlled. Another potential problem in PID tuning can occur with systems that have dual mode controls. These are systems that use one method to adjust the system in one direction and a different method to adjust it in the opposite direction. For example, consider a system in which we need the ability to quickly and accurately control the temperature of a liquid in either the positive or negative direction. Heating the liquid is a simple operation that could involve electric heaters. However, since hot liquids cool slowly, we must rapidly remove the heat using a water cooling system. In this case, not only must the controller regulate the amount of heat that is either injected into or extracted from the system, but it must also decide which control system to activate to achieve the desired results. This type of controller sometimes requires the use of two PIDs that are alternately switched on depending on whether the PV is above or below the SP.

Any designer who is familiar with the mathematical fundamentals of PID and the effect that each of the adjustments has on the system can eventually tune a PID to be stable and respond correctly (assuming the system can be tuned at all). However, to efficiently and quickly tune a PID, a designer needs the theoretical knowledge of how a PID functions, a thorough familiarity with how the particular machine being tuned responds to CV inputs, and experience at tuning PIDs.

It seems as though every designer with experience in tuning PIDs has his or her own personal way of performing the tuning. However, there are two fundamental methods that can be best used by someone who is new to and unfamiliar with PID tuning. As with nearly all PID tuning methods, both of these methods will give "ballpark" results. That is, they will allow the designer to "rough" tune the PID so that the machine will be stable and will function. Then, from this point, the PID parameters may be further adjusted to achieve results that are closer to the desired performance. There is no PID tuning method that will give the exact desired results on the first try (unless the designer is very lucky!).

Theoretically, it is possible to mathematically calculate the PID coefficients and accurately predict the machine's performance as a result of the PID tuning; however, in order to do this, the transfer function of the machine being controlled must be accurately mathematically modeled. Modeling the transfer function of a large machine requires determining mechanical parameters (such as mass, friction, damping factors, inertia, windage, and spring constants) and electrical parameters (such as inductance, capacitance, resistance, and power factor), many of which are extremely difficult, if not impossible, to determine. Therefore, most designers forego this step and simply tune the PID using somewhat of a trial and error method.

10–8 The "Adjust and Observe" Tuning Method

As the name implies, the "adjust and observe" tuning method involves the making of initial adjustments to the PID constants, observing the response of the machine, and then, knowing how each of the functions of the PID performs, making additional adjustments to correct for undesirable properties of the machine's response. From our previous discussion of PID performance, we know the following characteristics of PID adjustments:

1. Increasing the proportional gain k_p will result in a faster response and will reduce (but not eliminate) offset. However, at the same time, increasing proportional gain will also cause overshoot, hunting, and possible oscillation.

2. Increasing the derivative time constant T_d will reduce the hunting and overshoot caused by increasing the proportional gain. However, it will not correct for offset.

3. Decreasing the integral time constant T_i (also called the reset rate or reset time constant) will cause the PID to reduce the offset to near zero. Smaller values of T_i will cause the PID to eliminate the offset at a faster rate. Excessively small (non-zero) values of T_i will cause integral oscillation.

The adjustment procedure is as follows:

1. Initialize the PID constants. This is done by disabling the derivative and integral functions by setting both T_d and T_i to zero and setting k_p to an initial value between 1 and 5.

2. With the machine operating, quickly move the setpoint to a new value and observe the response. Figure 10–17 shows some typical responses with a step setpoint change from zero to 10 for various values of k_p. As shown in the figure, a good preliminary adjustment of k_p will result in a response with an overshoot that is approximately 10–30 percent of the setpoint change. At this point in the adjustment process we are not concerned about the minor amount of hunting, nor the offset. These problems will be corrected later by adjusting T_d and T_i respectively.

There is usually a large range that k_p can have for this adjustment. For example, the three responses shown in Figure 10–17 use k_p values of 2, 20, and 200 for this particular system. Although $k_p = 20$ may not be the final value we will use, we will leave it at this value for a starting point.

FIGURE 10–17

*k_p Adjustment
Responses*

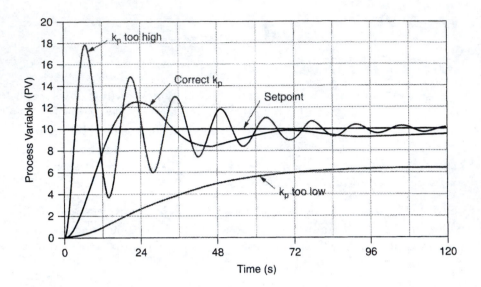

3. Increase T_d until the overshoot is reduced to a desired level. If no overshoot is desired, this can also be achieved by further increases in T_d. Figure 10–18 shows our example system with $k_p = 20$ and several trial values of T_d. For our system, we will attempt to tune the PID to provide minimal overshoot. Therefore, we will use a value of $T_d = 8$.

FIGURE 10–18

T_d Adjustment

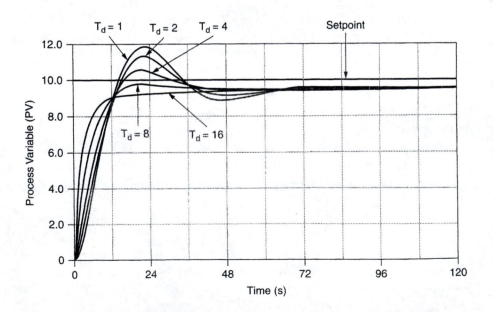

4. Adjust T_i such that the PID will eliminate the offset. Since T_i is the inverse of k_i ($T_i = 1/k_i$), this adjustment should begin with high values of T_i and then be reduced to achieve the desired response. For this adjustment, T_i values that are too large will cause the system to be slow in eliminating the offset. Values of T_i that are too small will cause the PID system to correct the offset too quickly and it will tend to oscillate. Figure 10–19 shows our example system with $k_p = 20$, $T_d = 8$, and values of 1000, 100, and 10 for T_i. If we are tuning the system for minimal overshoot, a value of $T_i = 100$ is a good choice.

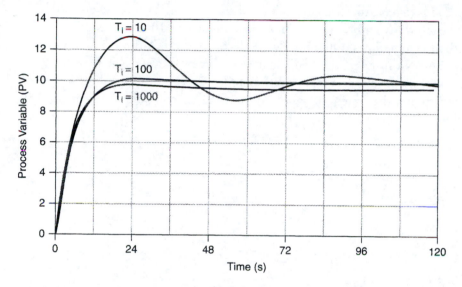

FIGURE 10–19

T_i Adjustment

5. Once the initial PID tuning is complete, the designer may now make further adjustments to the three tuning constants, if desired. From this starting point, the designer has the option to vary the proportional gain k_p over a wide range. This can be done to obtain a faster response to a setpoint change. If the system is to operate with varying loads, it is extremely important to test it for system stability under all load conditions.

10-9 The Ziegler-Nichols Tuning Method

The Ziegler-Nichols Tuning Method[3] (also called the ZN method) was developed in 1942 by two employees of Taylor Instrument Companies of

---

[3] J. G. Ziegler and N. B. Nichols, "Optimum Setting for Automatic Controllers," Trans. ASME, Vol. 64, 759–768, Nov. 1942.

Rochester, New York. J. G. Ziegler and N. B. Nichols proposed that consistent and approximate tuning of any closed-loop PID control system could be achieved by a mathematical process that involves measuring the response of the system to a change in the setpoint and then performing a few simple calculations. This tuning method results in an overshoot response, which is acceptable for many control systems and at least a good starting point for the others. If the desired response is to have less overshoot or no overshoot, some additional tuning will be required. The target amount of overshoot for the ZN method is to have the peak-to-peak amplitude of each cycle of overshoot be ¼ of the previous amplitude, as illustrated in Figure 10–20. Hence, the ZN method is sometimes called the ¼ wave decay method. Although it is unlikely that the ZN tuning method will achieve an exact ¼ wave decay in the system response, the results will be a stable system, and the tuning will be approximated so that the system designer can do the final tweaking using an "adjust and observe" method.

FIGURE 10–20

¼ Wave Decay

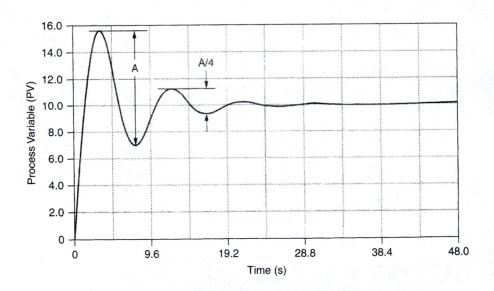

The main advantage in using the ZN tuning method is that all three tuning constants, k_p, T_d, and T_i, are pre-calculated and input to the system at the same time. It does not require any trial and error to achieve initial tuning. The chief disadvantage in using ZN tuning is that in order to obtain the system's characteristic data to make the calculations, the system must be operated either closed-loop in an oscillating condition or open loop. We shall investigate both approaches.

Oscillation Method

Using the oscillation method, we will be determining two machine parameters called the ultimate gain k_u and the ultimate period T_u. Using these, we will calculate k_d, T_i, and T_d.

1. Initialize all PID constants to zero. Power-up the machine and the closed-loop control system.

2. Increase the proportional gain k_p to the minimum value that will cause the system to oscillate. This must be a sustained oscillation (i.e., the amplitude of the oscillation must be neither increasing nor decreasing). It may be necessary to make changes in the setpoint to induce oscillation.

3. Record the value of k_p as the ultimate gain k_u.

4. Measure the period of the oscillation waveform. The period is the time (in seconds) for it to complete one cycle of oscillation. This period is the ultimate period T_u.

5. Shut down the system and readjust the PID constants to the following values:

$$k_p = 0.6\ k_u$$
$$T_i = 0.5\ T_u$$
$$T_d = 0.125\ T_u$$

As an example, we will apply the Ziegler-Nichols oscillation tuning method to our example motor speed control system that was used earlier in this chapter. First, with all of the gains set to zero, we increase the proportional gain k_p to the point where the process variable is a sustained oscillation. This is shown in Figure 10–21 and occurs at $k_p = 505$ (which is the ultimate gain k_u).

It is determined from this graph that, because it makes ten cycles of oscillation in 90 seconds, the period of oscillation and the ultimate period T_u is approximately 9 seconds. Next, we use the previously defined equations to calculate the three gain constants.

$$k_p = 0.6\ k_u = (0.6)(505) = 303$$
$$T_i = 0.5\ T_u = (0.5)(9) = 4.5$$
$$T_d = 0.125\ T_u = (0.125)(9) = 1.125$$

FIGURE 10–21

*ZN Tuning
Method Oscillation*

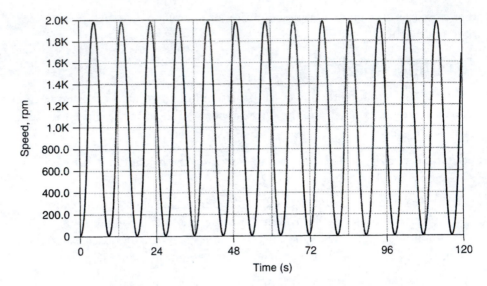

We then input these gain constants into the system and test the response with a
setpoint change from zero to 1,000, which is shown in Figure 10–22.

FIGURE 10–22

*Final Result
of ZN Tuning,
Oscillation
Method*

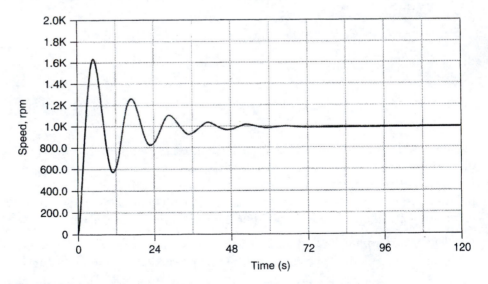

Open-Loop Method

For the open-loop method, we will be determining three machine parameters
called the deadtime L, the process time constant T, and the process gain K. Using
these, we will calculate k_d, T_i, and T_d. These measurements will be made with the

system in an open-loop unity gain configuration; that is, with the process variable PV (the feedback) open circuited, the proportional gain set to 1 ($k_d = 1$), and the integral and derivative gains set to zero ($T_i = T_d = 0$). The tuning steps are as follows.

1. Input a known change in the setpoint. This should be a step change.

2. Using a chart recorder, record a graph of the process variable PV with respect to time (with time zero being the instant that the step input is applied).

3. On the graph, draw three intersecting lines. First, draw a line that is tangential to the steepest part of the rising waveform. Second, draw a horizontal line on the left of the graph at an amplitude equal to the initial value of the PV until it intersects the first line. Third, draw a horizontal line on the right of the graph at an amplitude equal to the final value of the PV until it intersects the first line.

4. Find the deadtime L as the time from zero to the point where the first and second lines intersect.

5. Find the process time constant T as the time difference between the points where the first line intersects the second and third.

6. Find the process gain K as the percent of the PV with respect to the CV (the PV divided by the CV).

7. Shut down the system and readjust the PID constants to the following values:

$$k_p = 1.5T/(KL)$$

$$T_i = 2.5L$$

$$T_d = 0.4L$$

As an example, we will apply the Ziegler-Nichols open-loop tuning method to the motor speed control system that was shown earlier in this chapter. Figure 10–23 shows the open-loop response of the system with a SP change of zero to 1,000.

Next, we will add the three specified lines to the graph as shown in Figure 10–24.

From the graph we can see that the deadband L is approximately 7 seconds, the process time constant T is 135 seconds, and the change in the PV is 980

FIGURE 10–23

ZN Tuning Open Loop Response

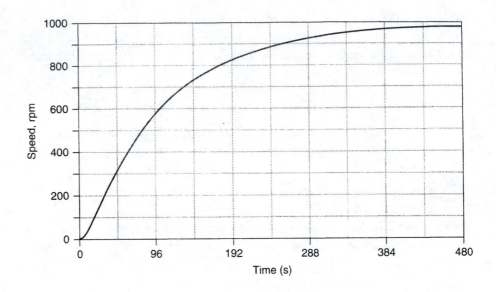

FIGURE 10–24

System Open-Loop Step Response

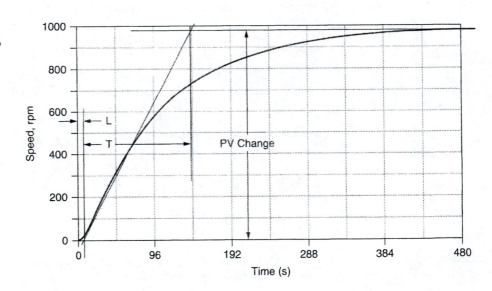

(98 percent of the SP), which results in a process gain K of 0.98. Next, we substitute these constants into our previous equations:

$$k_p = 1.5T/(KL) = (1.5 \times 135)/(0.98 \times 7) = 29.5$$

$$T_i = 2.5L = (2.5)(7) = 17.5$$

$$T_d = 0.4L = (0.4)(7) = 2.8$$

When these PID constants are loaded into the controller and the system is tested with a zero to 1,000 rpm step setpoint input, it responds as shown in Figure 10–25.

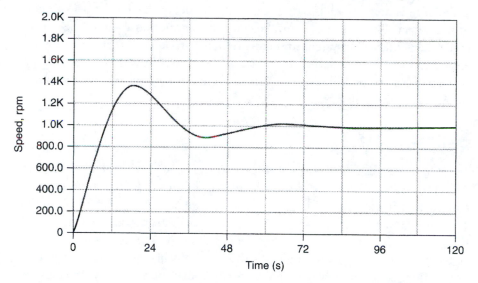

FIGURE 10–25

ZN Tuning Result Using the Open-Loop Method

Comparing the system step responses shown in Figures 10–22 and 10–25, we can see that, although they are somewhat different in the response time and the amount of hunting, both systems are stable and can be further tuned to the desired response. It should be stressed that the Ziegler-Nichols tuning method is not intended to allow the designer to arrive at the final tuning constants, but instead to provide a stable starting point for further fine tuning of the system.

10–10 Autotuning PID Systems

Some of the more sophisticated PLCs have a PID **autotune** feature that allows the PID to automatically determine the optimum k_d, T_d, and T_i values for a particular system. Although each manufacturer's autotune system has unique characteristics, the autotune procedure generally involves allowing the PID controller to output known CV values (usually step functions) to the controlled system and automatically analyzing the resulting transient response in the PV. Then, by using mathematical algorithms, the controller calculates the optimum k_d, T_d, and T_i values and stores them for use by the PID controller. The autotune procedure can take from a few seconds to many minutes to perform depending on the response time of the system being controlled and is usually manually initiated by maintenance or engineering personnel.

There are available controllers that carry the autotune feature to a higher level by performing continuous autotuning. For these controllers, the programmer inputs a **model reference,** which is basically a digital model of the desired performance of the system. Then, in operation, the controller continuously tunes the control constants of the actual system so that its response tracks that of the model reference. Model reference controllers are useful on systems in which the system parameters (such as mass, inertia, friction, etc.) may change during operation or if the system is non-linear.

Summary

As we have seen, tuning a PID control system can be somewhat of a "trial and error" exercise, usually because of the difficulty encountered if we attempt to mathematically characterize a machine system so that we can calculate the exact tuning constants. However, by knowing the effect that each of the tuning constants has on the system performance, we can use this insight to make the tuning task easier. In addition, the Ziegler-Nichols tuning methods can quickly provide us with starting point tuning constants. Then, with the system "rough tuned," the system designer can concentrate on the fine-tuning process to achieve the desired performance.

Review Questions

1. Why is proportional control alone usually insufficient to control most systems?

2. In a closed-loop system the process variable PV is 5.6 volts and the setpoint SP is 3.7 volts. What is the error voltage?

3. On graph paper, sketch the resulting waveform when taking derivative of a 1 Hz triangular waveform with a peak-to-peak amplitude of 2 volts (i.e., ± 1 volt).

4. What is the main effect that derivative gain K_d has on a closed-loop PID system?

5. Cite one disadvantage in using excessive derivative gain K_d.

6. Explain why the derivative function in a PID control system is bandwidth limited.

7. On graph paper, sketch the resulting waveform when taking the integral of a 3 volt dc voltage that is switched on from time $t = 1$ second until $t = 3$ seconds. Plot your graph from $t = 0$ to $t = 5$ seconds.

8. What is the main effect that integral gain K_i has on a closed-loop PID system?

9. Cite one disadvantage in using excessive integral gain K_i.

10. When a closed-loop system is oscillating, the process variable PV is a 5 Hz sine wave. What would be the ultimate period?

11. The proportional gain k_p of a control system is increased until the system exhibits a sustained oscillation. At this point, the proportional gain k_p is 325 and the frequency of oscillation is 0.25 Hz. Using the Ziegler-Nichols oscillation method, calculate the appropriate values of the PID constants k_p, T_i, and T_d.

12. A PID temperature control system is to be tuned using the Ziegler-Nichols open-loop method. The PV signal is disconnected, k_p is set to 1, and T_i and T_d are set to zero. A step input of 300° is input to the SP input and the PV shown in Figure 10–26 is recorded. Determine the PID constants k_p, T_i, and T_d.

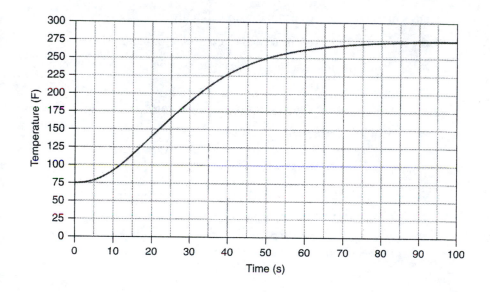

FIGURE 10–26

Graph for Problem 12

Motor Controls

Objectives

Upon completion of this chapter, you will know:

- why a motor starter is needed to control large ac motors.

- the components that make up a motor starter and how it operates.

- why motor overload protection on ac motors is needed and how a eutectic metallic alloy motor overload operates.

- why, until recently, ac motors were used for constant speed applications, and dc motors were used for variable speed applications.

- how a pulse width modulated (PWM) dc motor speed control controls dc motor speed.

- how a variable frequency motor drive (VFD) operates to control the speed of an ac induction motor.

Introduction

Since most heavy machinery is mechanically powered by electric motors, the system designer must be familiar with techniques for controlling electric motors. Motor controls cover a broad range from simple on-off motor starters to sophisticated phase-angle controlled dc motor controls and variable frequency ac motor drive systems. In this chapter, we will investigate some of the more common and popular ways of controlling motors. Since most heavy machinery requires large horsepower ac motors, and since large single-phase motors are not economical, our coverage of ac motor controls will be restricted to three-phase systems only.

11–1 AC Motor Starter

In its simplest form, a motor starter performs two basic functions. First, it allows the machine control circuitry (which is low voltage, low current, and either dc or single-phase ac) to control a high current, high voltage, multi-phase motor. This isolates the dangerous high voltage portions of the machine circuits from the safer low voltage control circuits. Second, it prevents the motor from automatically starting (or resuming) when power is applied to the machine, even if power is removed for a very short interval. Motor starters are commercially available devices. At a minimum, they include a relay (in this case it is called a contactor) with three heavy-duty N/O **main contacts** to control the motor, one light duty N/O **auxiliary contact** that is used in the control circuitry, and one light duty N/C **overload contact** that opens if a current overload condition occurs. This is shown in Figure 11–1.

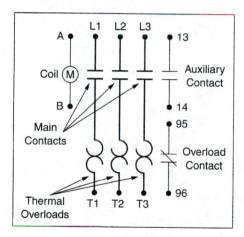

FIGURE 11–1

Simple Three-Phase Motor Starter with Overloads

Most starters have terminal labels (letters or numbers) etched or molded into the body of the starter next to each screw terminal that the designer may reference on schematic diagrams as shown in Figure 11–1. The coil of the contactor actuates all of the main contacts and the auxiliary contact at the same time. The overload contact is independent of the contactor coil and only operates under an overload condition. More complex starters may have four or more main contacts (instead of three) to accommodate other motor wiring configurations, and extra auxiliary and overload contacts of either the N/O or N/C type. In most cases, extra auxiliary and overload contacts may be added later as needed by "piggy-backing" them onto existing contacts.

Figure 11–2 illustrates one method of connecting the starter into the machine control circuitry. The terminal numbers on this schematic correspond to those on the motor starter schematic in Figure 11–1. Power from the three-phase line (sometimes called the **mains**) is applied to terminals L1, L2, and L3 of the starter, while the motor to be controlled is connected to terminals T1, T2, and T3.

FIGURE 11–2

Typical Motor Starter Application

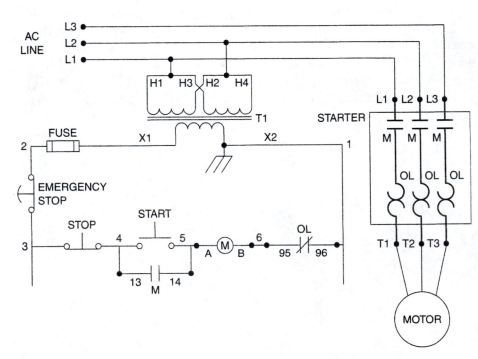

In operation, when the rails are powered, pressing the START switch provides power to the motor starter coil M. As long as the overload contacts OL are closed, the motor starter will actuate and three-phase power will be provided to the motor. When the starter actuates, auxiliary contact M closes, which bypasses the START switch. At this point, the START switch no longer needs to be pressed in order to keep the motor running. The motor will continue to run until (1) power fails, (2) the EMERGENCY STOP button is pressed, (3) the STOP switch is pressed, or (4) the overload contact OL opens. When any one of these four events occurs, the motor starter coil M de-energizes, the three-phase line to the motor is interrupted, and the auxiliary contact M opens.

11–2 AC Motor Overload Protection

For most applications, ac induction motors are overload protected at their rated current. Rated current is the current in each phase of the supplying line when operating at rated load and is always listed on the motor nameplate. Overload protection is required to prevent damage to the motor and feed circuits in the event a fault condition occurs, which includes a blocked rotor, rotor stall, and internal electrical faults. In general, simple single-phase fuses are not used for motor overload protection. When a motor is started, the starting currents can range from five to fifteen times the rated full-load current. Therefore, a fuse that is sized for rated current would blow when the motor is started. Even worse, since we would need to fuse each of the three phases powering a motor, if only one of the fuses were to blow, the motor would go into what is termed a **single-phasing condition.** In this case, the motor shaft may continue to rotate (depending on the mechanical load), but the motor will operate at a drastically reduced efficiency causing it to overheat and eventually fail. Therefore, the motor overload protection must (1) ignore short term excessive currents that occur during motor starting and (2) simultaneously interrupt all three phases when an overload condition occurs.

The solution to the potential single-phasing problem is to connect a current-sensing device (called an **overload**) in series with each of the three phases and to mechanically link them so that when any one of the three overloads senses an over-current condition, it opens a contact (called an overload contact). The overload contact is connected into the motor starter circuit so that when the overload contact opens, the entire starter circuit is disabled, which in turn opens the three-phase motor contactor interrupting all three phases powering the motor. The nuisance tripping problem is overcome by designing the overloads to react slowly and, therefore, not trip when a short-term overload occurs, such as the normal starting of the motor.

The most popular type of overload is the thermal overload. Although some thermal overloads use a bi-metallic temperature switch (the bi-metallic switch was covered earlier in this text), the more popular type of thermal overload is the **eutectic metallic alloy overload.** This device, shown in Figure 11–3, consists of a **eutectic alloy,**[1] which is heated by an electrical coil (called a **heater**)

---

[1] A eutectic alloy is a combination of metals that has a very low melting temperature and changes quickly from the solid to liquid state, much like solder, instead of going through a "mushy" condition during the state change.

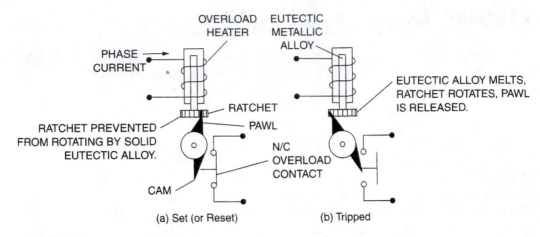

OVERLOAD   EUTECTIC
HEATER   METALLIC
ALLOY

PHASE
CURRENT

EUTECTIC ALLOY MELTS,
RATCHET ROTATES, PAWL
IS RELEASED.

RATCHET

RATCHET PREVENTED
FROM ROTATING BY SOLID
EUTECTIC ALLOY.

PAWL

N/C
OVERLOAD
CONTACT

CAM

(a) Set (or Reset)   (b) Tripped

FIGURE 11–3

Eutectic Metallic Alloy Thermal Overload (One Phase)

through which the phase current passes. If the phase current through the overload heater is excessive, the eutectic metal will eventually melt. This releases a ratchet, which cams the normally closed overload contact into the open position. In order to make the overload reusable, the eutectic alloy in the overload is sealed in a tube so that it will not leak out when it melts. The sealed tube, eutectic metallic alloy, and spindle are commonly called the **overload spindle,** which is illustrated in Figure 11–4

FIGURE 11–4

*Overload Heater
and Spindle*

When the overload cools and the eutectic alloy solidifies, the ratchet again will be prevented from rotating and the overload can then be reset by manually pressing a reset lever. For clarity, only one phase of the overload system is shown in Figure 11–3. In practice, for three-phase systems there are three overloads operating the same overload contact.

One major advantage in using the thermal overload is that, as long as the overload is properly sized for the motor, the overload heats in much the same manner as the motor itself. Therefore, the temperature of the overload is a good indicator of the temperature of the motor windings. Of course, this principle is what makes the overload a good protection device for the motor. However, for this reason, the designer must consider any ambient temperature difference between the motor and the overload. If the motor and overload cannot be located in the same area, the overload size must be readjusted using a temperature correction factor. It would be unwise to locate a motor outdoors but install the overload indoors in an area that is either heated in the winter or cooled in the summer without applying a temperature correction factor when selecting the overload.

Another advantage in using this type of overload is that the starter can be resized to a different trip current by simply changing the overload. A variety of off-the-shelf overloads are available, and they can be easily changed in a few minutes using a screwdriver.

11-3 Specifying a Motor Starter

Motor starters must be selected according to the (1) number of phases to be controlled, (2) motor size, (3) coil voltage, and (4) overload heater size. Additionally, the designer must specify any desired optional equipment such as the number of auxiliary contacts and overload contacts and their form types (form A or form B) and whether the starter is to be reversible. When selecting a starter, most system designers simply choose a starter manufacturer, obtain their catalog, and follow the selection guidelines in the catalog.

1. Most motor starter manufacturers specify the contact rating by motor horsepower and line voltage. For most squirrel cage induction motors, knowing these two values will completely define the amount of voltage and current that the contacts must switch.

2. By knowing the full-load motor current (from the motor's nameplate), the overload heater can be selected. Because the heater is a resistive type heater, the heat produced is a function of only the heater resistance and the phase current.

Line voltage has no bearing on heater selection. Since the overload spindle heats slowly, normal starting current or short duration over-current conditions will not cause the overload to trip so it is not necessary to oversize the heater.

3. Since the contactor actuating coil is operated from the machine control circuitry, the coil voltage must be the same as that of the controls circuit, and either ac or dc operation must be specified. This may or may not be the same voltage as the contact rating and must be specified when purchasing the contactor. Contactors with ac and dc coils are not interchangeable even if their rated voltages are the same.

4. Auxiliary contacts and additional overload contacts are generally mounted to the front or side of the contactor. The designer may specify form A or form B types for either of these contacts.

11-4 DC Motor Controller

Before the advent of modern solid-state motor controllers, most motor applications that required variable speed required the use of dc motors. The reason is that the speed of a dc motor can be easily controlled by simply varying either the voltage applied to the armature or the current in the field winding. In contrast, the speed of an ac motor is determined, for the most part, by the frequency of the applied ac voltage. Since electricity from the power company is delivered at a fixed frequency, ac motors were relegated to constant speed applications (such as manufacturing machinery), while dc motors were used whenever variable speed was required (e.g., in cranes and elevators). This situation changed radically when variable frequency solid-state motor controllers were introduced, which now allow us to easily and inexpensively operate ac motors at variable speeds.

Although solid-state controllers have allowed ac motors to make many inroads into applications traditionally done by dc motors, there are still many applications where a dc motor is the best choice for the job, mostly in battery powered applications. These are usually in the areas of small instrument motors used in robotics, remote-controlled vehicles, toys, battery operated electro-mechanical devices, automotive applications, and spacecraft. Additionally, the universal ac motor, which is used in ac-operated power hand tools and kitchen appliances, is actually a spinoff of the series dc motor design (it is not an induction motor) and can be controlled in the same manner as a dc motor (i.e., by varying the applied voltage).

The voltage applied to the armature and the current through the field of a dc motor can be reduced by simply adding a resistance in series with the armature or field, which is a lossy and inefficient method. Under heavy mechanical loads, the high armature currents cause an exponentially high I^2R power loss in any resistor in series with the armature. Solid-state electronics has made improvements in the control of dc motors in much the same way that it has with ac motors. In this chapter, we will examine two of these solid-state control techniques. In the first, we will be controlling the speed of a dc motor that is powered from a dc source. In the second, we will be using an ac power source. In both cases, we will control the motor speed by varying the average armature voltage. We will assume the field is held constant either by the use of permanent magnets or by a constant field current.

dc Motor Control with dc Power Source

Consider the circuit shown in Figure 11–5. In this case, we have a dc voltage source V, a resistor R, inductor L, diode D, and a semiconductor switch Q (shown here as an N-channel insulated gate MOSFET). The signal applied to the gate of the switch Q is a pulse train with constant frequency f (and constant period T), but with varying pulse width t. The amplitude of the signal applied to the gate will cause the switch to transition between cutoff and saturation with very short rise and fall times. The relative values of R and L are selected such that the time constant $\tau = L/R$ is at least 10 times the period T of the pulse train applied to the gate of Q. The long L/R time constant will have a low-pass filtering effect on the chopped output of the switch Q, and will effectively smooth the current into dc with very little ac component.

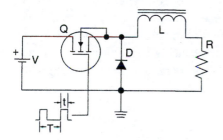

FIGURE 11–5

Simple dc Switch Voltage Controller

For the switch Q, the ratio of the on-time t to the period T is defined as the **duty cycle** and is represented as a percentage between 0 percent and 100 percent. If we apply a pulse train with a 0 percent duty cycle to the gate of the switch, the switch will remain off all of the time and the voltage on the resistor R will obviously be zero. In a similar manner, if we apply a pulse train with a duty cycle of 100 percent, the switch will remain on all the time, the diode D will be reverse biased, and, after five time constants, the voltage on the resistor R will be V. For any duty cycle between 0 percent and 100 percent, the *average* resistor voltage will be a corresponding percentage of the voltage V. For example, if we adjust the applied gate pulses so that the duty cycle is 35 percent (i.e., ON for 35 percent of the time, OFF for 65 percent of the time), then the voltage on the resistor R will be 35 percent of the input voltage V. This is because during the time that the switch is ON the inductor L will store energy; during the time the switch is OFF, the inductor will give up some of its stored energy keeping current flowing in the circuit through inductor L, resistor R, and the forward biased diode D (in this application, the diode is called a **freewheeling diode** or **commutation diode**). Although the operating frequency of the switch will affect the amplitude of the ac ripple voltage on R, it will not affect the average voltage. The average voltage is controlled by the duty cycle of the switch. Whenever the duty cycle is changed, the voltage on the resistor will settle within five L/R time constants.

Although this seems to be a rather involved method of controlling the voltage on a load, there is one major advantage in using this method. Consider the power dissipation of the switch during the times when it is either OFF or ON. The power dissipated by any component in a dc circuit is simply the voltage drop on the component times the current through the component, or V times I. Assuming an ideal switch, when the switch is OFF, the current will be zero, resulting in zero power dissipation. Similarly, when the switch is ON, the voltage drop will be near zero, which also results in near-zero power dissipation. However, when the switch changes state, there is a very short period of time when the voltage and current are simultaneously non-zero and the switch will dissipate power. For this reason, when constructing circuits of this type, the most important design considerations are the on-resistance of the switch, the off-state leakage current of the switch, the rise and fall times of the pulse train, and the speed at which the switch can change states (which is usually a function of the interelectrode capacitance within the switch). If these parameters are carefully controlled, it is not unusual for this circuit to have an efficiency that is in the high 90 percent range.

With this analysis in mind, consider the electrical model of the armature of a dc motor shown in Figure 11–6. Since the armature windings are made from copper wire, there will be distributed resistance in the coils R_a, and the coils themselves will give the armature distributed inductance L_a. Therefore, we generally model the armature as a lumped resistance and a lumped inductance connected in series. Some armature models include a brush and commutator voltage drop component (called **brush drop**), but for the purpose of this analysis, since the brush drop is relatively constant and small, adding this component will not affect the outcome.

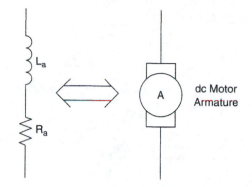

FIGURE 11–6

dc Motor
Armature Model

Since we can model the dc motor armature as a series resistance and inductance, we can substitute the armature in place of the resistor and inductor in our dc switch circuit in Figure 11–5. This modified circuit is shown in Figure 11–7.

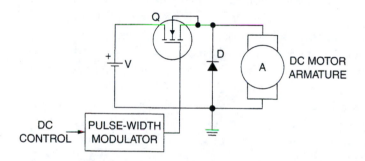

FIGURE 11–7

dc Motor Speed
Control

Note in Figure 11–7 that a pulse-width modulator has been added. This is a sub-system that converts a dc control input voltage to a constant-frequency variable duty cycle pulse train. The complexity of this function block depends on the sophistication of the circuit and can be as simple as a 555 integrated circuit timer or as complex as a single board microcontroller. Since the heart of this entire control circuit is the pulse-width modulator, the entire system is generally called a **pulse-width modulator motor speed control.**

Although the circuit in Figure 11–7 illustrates how a dc motor can be controlled using a pulse-width modulator switch, there is a minor problem in that the switching transistor is connected as a source follower (common drain) amplifier. This circuit configuration does not switch as fast nor does it saturate as well as the grounded source configuration, which results in the transistor dissipating excessive power. Therefore, we will improve the circuit by exchanging the positions of the motor armature and the switch. This circuit is shown in Figure 11–8.

FIGURE 11–8

Improved Pulse-Width Modulator dc Motor Speed Controller

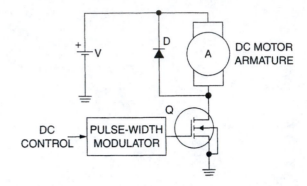

dc Motor Control with ac Power Source

Obviously, in order to operate a dc motor from an ac source, the ac power must first be converted to dc. It should be noted that the dc used to operate a dc motor need not be a continuous non-varying voltage. It is permissible to provide dc power in the form of half-wave rectified or full-wave rectified power or even pulses of power (as in the pulse-width modulation scheme discussed earlier). Generally speaking, as long as the polarity of the voltage applied to the armature does not reverse, the waveshape is not critical. In its simplest form, a full-wave rectifier circuit shown in Figure 11–9 will power a dc motor armature. However, in this circuit, the average dc voltage applied to the armature will be dependent on the ac line voltage. If we desire to control the speed of the motor, we would need the ability to vary the ac line voltage.

FIGURE 11–9

Full-Wave Rectifier dc Motor Supply

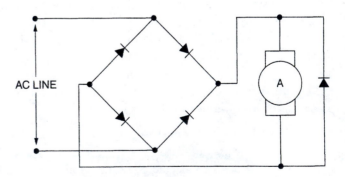

To provide control over the average dc voltage applied to the motor without the need to vary the line voltage, we will replace two of the rectifier diodes in our bridge with silicon controlled rectifiers (SCRs) as shown in Figure 11–10.

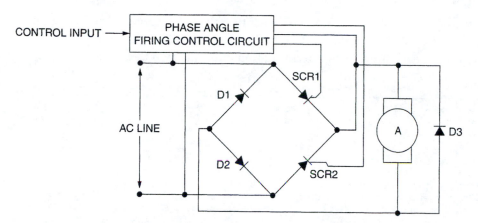

FIGURE 11–10

*SCR dc Motor
Control Circuit*

Note in the circuit in Figure 11–10 that it is not necessary for all four rectifier diodes to be SCRs. The reason is that the remaining two diodes, D1 and D2, are simply steering diodes that route the return current from the armature back to the opposite side of the ac line. Therefore, we can completely control current flow in the circuit with the two SCRs shown. The phase angle firing control circuit monitors the sine wave of the ac line and produces a precisely timed pulse to the gate of each of the SCRs. As we will see, this will ultimately control the magnitude of the average dc voltage applied to the motor armature.

We will begin analyzing our circuit using the two extreme modes of operation, which are when the SCRs are fully OFF and fully ON. First, assume that the firing control circuit produces no signal to the gates of the SCRs. In this case, the SCRs will never fire and there will be zero current and zero voltage applied to the motor armature. In the other extreme, assume that the firing control circuit provides a short pulse to the gate of each SCR at the instant that the anode-to-cathode voltage (V_{A-K}) on the SCR becomes positive (which occurs at 0° in the applied sine wave for SCR1 and 180° for SCR2). When this occurs, the SCRs will alternately fire and remain fired for their entire respective half cycle of the sine wave. In this case, the two SCRs will act as if they are rectifier diodes and the circuit will per- form the same as that in Figure 11–9 with the motor operating at full speed.

Now, consider the condition in which the SCRs are fired at some delayed time after their respective anode-to-cathode voltages (V_{A-K}) become positive. In this case, there will be a portion of the half-wave rectified sine wave missing from the output waveform, which is the portion from the time the waveform starts at zero until the time the SCR is fired. When we fire the SCRs at some delayed time, we generally measure this time delay as a trigonometric angle (called the **firing angle**) with respect to the positive-slope zero crossing of the sine wave of the line voltage. Figure 11–11 shows the armature voltage for various firing angles between 0° and 180°. Notice that as the SCR firing angle increases from 0° to 180°, the motor armature voltage decreases from full voltage to zero.

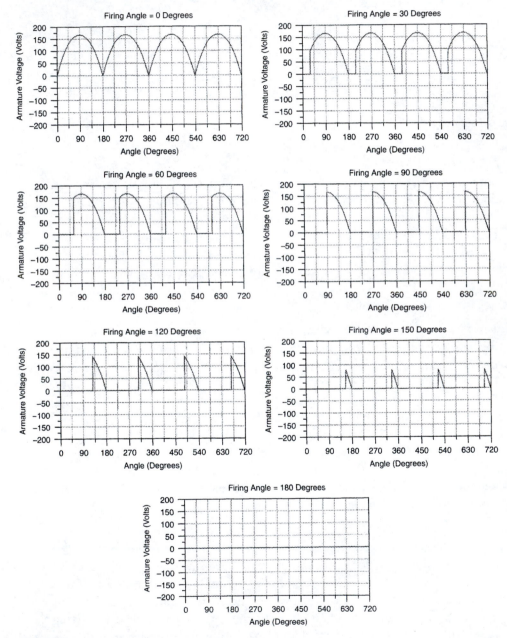

FIGURE 11–11

Armature Voltage for Various Firing Angles

The circuit in Figure 11–10 is designed to operate from single-phase power. For most small horsepower applications, single-phase power is sufficient to operate the motor. However, for large dc motors, three-phase power is more suitable because, for the same size motor, it will reduce the ac line current, which consequently reduces the wire size and power losses in the feeder circuits. The circuit shown in Figure 11–12 is a simplified three-phase SCR control and rectifier for a dc motor that uses the same SCR firing angle control technique to control the motor armature voltage.

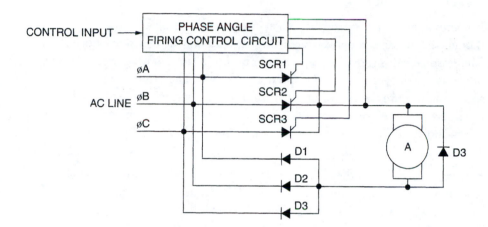

FIGURE 11–12
Three-Phase Powered dc Motor Speed Control

There are other more sophisticated techniques for controlling the speed of a dc motor, but most are variations on the two techniques covered. DC motor speed control systems are available as off-the-shelf units and are usually controlled by a dc voltage input, typically 0–10 volts dc, which can be easily provided by a PLC analog output.

11-5 Variable Speed (Variable Frequency) AC Motor Drive

As mentioned earlier, if we wish to control the speed of an ac induction motor and produce rated torque throughout the speed range, we must vary the frequency of the applied voltage. This sounds relatively simple; however, there is one underlying problem that affects this approach. For inductors (such as induction motors), as the frequency of an applied voltage is decreased, the magnetic flux increases. Therefore, as the frequency is decreased, if the voltage is maintained constant, the core of the inductor (the motor stator) will magnetically saturate, the line current will increase drastically, and it will overheat and fail. In order to maintain a constant flux, as we reduce the frequency, we must reduce the applied voltage by the same proportion. This technique is commonly applied when operating a 60 Hz induction motor on a 50 Hz line. The motor will operate safely and efficiently if we reduce the 50 Hz line voltage to 50/60 (or 83.3 percent) of the motor's rated nameplate voltage. This principle applies to the operation of an induction motor at any frequency; that is, the ratio of the frequency to line voltage f/V must be maintained constant. Therefore, if we wish to construct an electronic system to produce a varying frequency to control the speed of an induction motor, it must also be capable of varying its output voltage proportional to the output frequency.

It is important to recognize that when operating a motor at reduced speed, although the motor can deliver rated torque, it cannot deliver rated horsepower. The reason for this is that the horsepower output of any rotating machine is proportional to the product of the speed and torque. Therefore, if we reduce the speed and operate the motor at rated torque, the horsepower output will be reduced by the speed reduction ratio. Operating an induction motor at reduced speed and rated horsepower will cause excessive line current, overheating, and eventual failure of the motor.

Earlier, we investigated the technique of using a pulse-width modulation (PWM) technique to vary the dc voltage applied to the armature of a dc motor. Assume, for this discussion, we add a second pulse-width modulator but design it to produce negative pulses instead of positive pulses. We could connect the outputs of the two circuits in parallel to our load (a motor). By carefully controlling which of the two PWMs is operating at any given time and the duty cycle of each, we can produce any time-varying voltage with a peak voltage amplitude between +V and −V.

Such a device is capable of producing any wave shape of any frequency and of any peak-to-peak voltage amplitude (within the maximum voltage capability of the PWMs). Of course, for motor control applications, the desired waveshape will be a sine wave. Practically speaking, in order to do this, we would most likely need a microprocessor controlling the system that processes the desired frequency to obtain the correct line voltage required and the corresponding PWM duty cycles to produce a sine wave of the correct voltage amplitude and of the desired frequency.

The output waveform of such a device (called a **variable frequency drive**, or **VFD**) is shown in Figure 11–13. Superimposed on the pulse waveform is the desired sine wave. Notice that during the 0°–180° portion of the waveform, the positive voltage PWM is operating, and during the 180°–360° portion, the negative PWM is operating. Also notice that during each half cycle, the duty cycle starts at 0 percent, increases to nearly 100 percent, and then decreases to 0 percent. The duty cycle of each pulse is calculated and precisely timed by the PWM controller (the microprocessor) so that the average of the pulses approximates the desired sine wave.

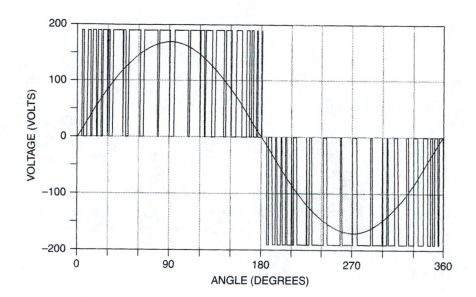

The amplitude of the voltage output of the VFD is controlled by proportionally adjusting the width of all the pulses. For example, if we were to reduce all of the pulses to 50 percent of their normal width, the average output waveform would still be a sine wave but would be 50 percent of the amplitude of the original waveform.

Since the VFD will be connected to an inductive device (an induction motor), the pulse waveform shown in Figure 11-13 is generally not viewable using an oscilloscope. The inductance of the motor will smooth the pulses in much the same manner as the dc motor smooths the PWM output, which, for the VFD, will result in voltage and current waveforms that are nearly sinusoidal.

It is important to recognize that the previous discussion is highly theoretical and oversimplified to make the concept more understandable. When a VFD is connected to an induction motor, there are many other non-ideal characteristics that are caused by the inductance of the motor and its magnetic characteristics, such as counter EMF, harmonics, energy storage, and power factor, which make the design of VFDs much more complex than can be comprehensively covered in this text. However, present day VFD technology utilizes the versatility of the internal microprocessor to overcome most of these adverse characteristics, making the selection and application of a VFD relatively easy for the end-user.

In addition to performing simple variable frequency speed control of an induction motor, most VFDs also provide a wealth of features that make the system more versatile and provide protection for the motor being controlled. These include features such as over-current monitoring, automatic adjustable overload trip, speed ramping, ramp shaping, rotation direction control, and dynamic braking. Some VFDs have the capability to operate from a single-phase line while providing three-phase power to the motor. Selection of a VFD usually requires simply knowing the line voltage, the motor's rated input voltage, and the horsepower.

Most VFDs can be controlled from a PLC by providing an analog (usually 0–10 volts) dc signal from the PLC to the VFD, which controls the VFD between zero and the rated frequency. Direction control is usually done using a discrete output from the PLC to the VFD. Although VFDs are very reliable, the designer should include a contactor to control the main power to the VFD. Because an electronic failure in the VFD or a line voltage disturbance can cause the VFD to operate unpredictably, it is unwise to allow the VFD to maintain the machine at rest while an operator accesses the moving parts of the machine.

Summary

It is important for the machine designer to be very familiar with various methods of controlling ac and dc motors. These range from the simple motor starter to the sophisticated pulse-width modulated (PWM) dc motor controls and the variable frequency (VFD) ac motor controllers. New advances in solid state power electronics have made the speed control of both ac and dc motors simple, efficient, and relatively inexpensive. The pulse-width modulator (PWM) system is capable of efficiently controlling the speed of a dc motor by controlling the average armature voltage of the motor. The variable frequency ac motor drive (VFD) is capable of controlling the speed of an ac induction motor by controlling both the frequency and amplitude of the applied three-phase power. As a result, the VFD has made it possible to use ac induction motors for variable speed applications that, until recently, were best performed by dc motors.

Review Questions

1. Explain the difference between a motor starter and a simple on-off switch.

2. For the circuit shown in Figure 11–2, assume the STOP switch is defective and remains closed all the time, even when it is pressed. How can the machine be stopped?

3. For the circuit shown in Figure 11–2, when the START switch is pressed, the motor starter contactor energizes as usual. However, when the START switch is released, the contactor de-energizes and the motor stops. What is the most likely cause of this problem?

4. A pulse-width modulation dc motor speed control is capable of 125 volts dc maximum output. What is the output voltage when the PWM is operating at 45 percent duty cycle?

5. For the PWM in problem 4, what duty cycle is required if the desired output voltage is 90 volts dc?

6. A PWM is operating at a frequency of 25 kHz. If it is outputting pulses at a 75 percent duty cycle, what is the width of each pulse in microseconds?

7. An ac induction motor is rated at 1,750 rpm with a line frequency of 60 Hz. If the motor is operated on a 50 Hz line, what will be its approximate speed?

8. An ac induction motor is rated at 1,175 rpm, 480 V, 60 Hz three-phase. If we reduce the motor speed by reducing the line frequency to 25 Hz, what should be the line voltage?

9. An induction motor is rated at 30 hp 1,175 rpm. If we connect the motor to a variable frequency ac drive and operate the motor at 900 rpm, what is the maximum horsepower the motor can safely deliver?

10. A VFD has an output frequency range of 0–120 Hz with an input control voltage of 0–10 volts dc. The VFD is connected to an induction motor that is rated at 875 rpm at 60 Hz. What dc input control voltage is needed to have the motor operate at 625 rpm?

System Integrity and Safety

Objectives

Upon completion of this chapter, you will know:

- how to assure that a system will be reliable in various environmental conditions.

- how to select a NEMA or IEC rated enclosure for electrical equipment.

- how to connect PLC inputs and outputs to be fail-safe.

- various ways to make a system more safe by installing interlocks.

Introduction

In addition to being able to design and program an efficient working control system, it is important for the system designer to be well aware of other non-electrical and non-software-related issues. These are issues that can cause the best and most clever designs to fail prematurely, work intermittently, or, even worse, to be a safety hazard. In this chapter, we will investigate some of the tools and procedures available to the designer so that the system will work well, work safely, and work with minimal down-time.

12-1 System Integrity

It is obvious that we would never consider exposing a twisted copper wire connection to the outdoor weather. Surely, the weather would eventually tarnish and corrode the connection, and the connection would either become intermittent or fail altogether. But what if the connection were outdoors, but under a roof—say a carport? Would a bare twisted wire connection be acceptable? And what if the same type of connection were used in a ceiling light fixture for an indoor swimming pool? Surely the chlorine used to purify the pool water will have some adverse effect on the twisted copper wires.

In general, how does a system designer know how to ward off environmental effects so that they will not cause premature failure of the system? One solution is to put all of the electrical equipment inside an enclosure or electrical box. But how do we know how well the electrical box will ward off the same environmental effects? Can we be sure it will not leak in a driving rainstorm? The answers lie in guidelines set forth by the National Electrical Manufacturers Association (NEMA) and the International Electrotechnical Commission (IEC) regarding electrical enclosures. NEMA is a United States based association, while IEC is based in Europe. Both have set forth similar standards by which manufacturers rate their products based on how impervious they are to environmental conditions. NEMA assigns a NEMA number to each classification, while IEC assigns an IP (Index of Protection) number. It is possible, to some extent, to be able to cross reference NEMA and IEC classes; however, there is not an exact one-to-one relationship between the two.

NEMA and IEC ratings are based mostly on the enclosure's ability to protect the equipment inside from accidental body contact, dust, splashing water, direct hosedown, rain, sleet, ice, oil, coolant, and corrosive agents. Since the designer knows the environment in which the equipment is to be used, it is relatively simple to look up the required protection in a NEMA or IEC table and then specify the appropriate NEMA or IP number when purchasing the equipment. Generally speaking, the NEMA and IP numbers are assigned so that the lower numbers provide the least protection while the highest numbers provide the best protection. Because of this, the cost of a NEMA or IEC rated enclosure is usually directly proportional to the NEMA or IP number.

Consider the NEMA enclosure ratings shown in Table 12–1. If, for example, we needed an enclosure to protect equipment from usual outdoor weather conditions, then a NEMA 3 or NEMA 4 enclosure would be acceptable. However, if the enclosure were near the ocean or a swimming pool where it would be exposed to corrosive salt water or chlorinated water splash, then a NEMA 4X would be a better choice. In a similar manner, we can conclude that underwater equipment must be NEMA 6P rated, and that an enclosure that is to be mounted on a hydraulic pump should be NEMA 12 or NEMA 13.

It would seem logical to simply use NEMA 6P for everything (except oil and coolant exposure). However, the very high cost of NEMA 6P enclosures prohibits their use in non-submerged applications. Therefore, because of cost constraints, it is also important to avoid overspecifying a NEMA enclosure.

Table 12–1 NEMA Enclosure Rating Table

NEMA #	1	2	3	3S	4	4X	6	6P	12	13
Suggested Usage (I = Indoor, O = Outdoor)	I	I	O	O	I/O	I/O	I/O	I/O	I	I
Accidental Body Contact	X	X	X	X	X	X	X	X	X	X
Falling Dirt	X	X	X	X	X	X	X	X	X	X
Dust, Lint, Fibers (non-volatile)			X	X	X	X	X	X	X	X
Windblown Dust			X	X	X	X	X	X		
Falling Liquid, Light Splash		X	X	X	X	X	X	X	X	X
Hosedown, Heavy Splash					X	X	X	X		
Rain, Snow, Sleet			X	X	X	X	X	X		
Ice Buildup				X						
Oil or Coolant Seepage									X	X
Oil or Coolant Spray or Wash										X
Occasional Submersion							X	X		
Prolonged Submersion								X		
Corrosive Agents						X		X		

IEC enclosure numbers address environmental issues as do the NEMA categories. However, the IEC numbers also address safety issues. In particular, they specify the amount of personal protection the enclosure offers in keeping out intrusion by foreign bodies such as hands, fingers, tools, and screws. IP numbers are always two-digit numbers. The leftmost digit (tens) specifies the protection against intrusion by foreign bodies while the rightmost digit (units) specifies the environmental protection provided by the enclosure. The IEC IP number ratings are shown in Table 12–2.

There is also a rating of IPx8, which is waterproof. There is no tens digit on this rating because, since it is waterproof, it is also naturally impervious to any and all foreign objects. Also, since there is only one column 7 rating (which is IP67), it is referred to as either IP67 or IPx7. In the IP_2 column, "inclined water" refers to rain or drip up to 15° from vertical, and in the IP_3 column, spray water can be up to 60° from vertical.

As some examples of how to use the IEC table, assume we wish to have an enclosure that will keep out rain (inclined water) and will not allow tools to be pushed into any openings. Locating those items in the columns and rows, we find

Table 12-2 IEC Enclosure Rating Table

		IEC IP Enclosure Ratings							
		No Protection	*Vert. Water*	*Inclined Water*	*Spray Water*	*Splash Water*	*Hose*	*Flooding*	*Dripping*
		IP_0	*IP_1*	*IP_2*	*IP_3*	*IP_4*	*IP_5*	*IP_6*	*IP_7*
No Protection	IPO_	X							
Foreign Obj. 50mm max (hand)	IP1_	X	X	X					
Foreign Obj. 12.5mm max (finger)	IP2_	X	X	X	X				
Foreign Obj. 2.5mm max (tools)	IP3_	X	X	X	X	X			
Foreign Obj. 1mm max (screws, nails)	IP4_	X	X	X	X	X			
Dust Protected	IP5_	X	X	X	X	X	X	X	
Dust Tight	IP6_	X	X	X	X	X	X	X	X

that an IP32 enclosure is needed. Additionally, an enclosure that will keep out hands and offers no environmental protection is an IP10 enclosure. Most consumer electronics products (stereos, televisions, VCRs, etc.) are IP40.

12-2 Equipment Temperature Considerations

It is a proven fact that the length of life of an electronic device is inversely proportional to the temperature at which it is operated. In other words, to make electronic equipment last longer, it should be operated in a low temperature environment. Obviously, it is impractical to refrigerate controls installations. However, it is important to take necessary steps to assure that the equipment does not overheat and that it does not exceed manufacturers' specifications of maximum allowable operating temperature.

When electrical equipment is installed inside a NEMA or IEC enclosure, it will most certainly produce heat when powered. This will raise the temperature inside the enclosure. It is important that this heat be somehow dissipated. Since most cabinets used in an often dirty manufacturing environment are sealed (to keep out dirt and dust), the most popular way to do this is to use the cabinet itself as a heat sink. Generally, the cabinet is made of steel and is bolted to a beam or to the metal side of the machine, which improves the heat sinking capability of the enclosure. If this type of enclosure mounting is not available, the temperature of the inside of the enclosure should be measured under worst-case conditions; that is, with all equipment in the enclosure operating under worst-case load conditions. If the temperature is excessive, other cooling methods must be considered.

Another way of controlling temperature inside an enclosure is by using a cooling fan. However, this will require screens and filters to cleanse the air being drawn into the cabinet. This, in turn, increases periodic maintenance to clean the screens and filters.

12-3 Fail-Safe Wiring and Programming

In most examples in the earlier chapters of this text, we used all normally open momentary pushbutton switches connected to PLCs. However, what would happen if we used a normally open pushbutton switch for a STOP switch and one of the wires on the switch became loose or broken, as shown in Figure 12–1? Naturally, when we press the STOP switch, the PLC will not receive a signal input, and the machine will simply continue running.

FIGURE 12–1

Non-Fail-Safe Wiring of STOP Switch

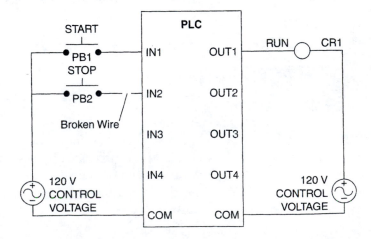

The problem is that a design of this type is not **fail-safe**. *Fail-safe design is a method of designing control systems such that if a critical component in the system fails, the system immediately becomes disabled.*

Let us reconsider our STOP switch example, except this time, we will have the STOP switch provide a signal to the PLC when we *do not* want it to stop. In other words, we will use a normally closed (N/C) pushbutton switch that, when pressed, will *break* the circuit. This change will also require that we invert the STOP switch signal in the PLC ladder program. Then, if one of the wires on the STOP switch breaks as shown in Figure 12–2, the PLC no longer "sees" an input from the STOP switch. The PLC will interpret this as if someone has pressed the switch, and it will stop the machine. In addition, as long as the PLC program is written such that the STOP overrides the START, then if the wire on the STOP switch breaks, not only will the machine stop, but pressing the START switch will have no effect either.

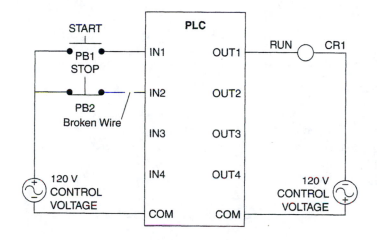

FIGURE 12–2

Fail-Safe Wiring of STOP Switch

Fail-safe wiring also applies to PLC outputs. Consider an application where a PLC is to control a crane. Naturally, there will be a disk braking system that will lock the wench and prevent a load on the crane from being lowered. In order to be fail-safe, this braking system needs to be ON when electrical power is OFF. In other words, it needs to be held ON by mechanical spring pressure and released by electrical, hydraulic, pneumatic, or any other method. In doing so, any failure of the powering system will cause the braking system to lose power, and the spring will automatically apply the brake. This means that it requires a relay contact closure from the PLC output to *release* the brake instead of *applying* the brake.

Since emergency stop switches are critical system components, it is important that these always operate correctly and that they are not buffered by some other electronic system. Emergency stop switches are always connected in series with the power line of the control system and, when pressed, will interrupt power to the controls. When this happens, fail-safe output design will handle the disabling and halting of the system.

Since PLC ladder programming is simply an extension of hard wiring, it is important to consider fail-safe wiring when programming also. Consider the start/stop program rung shown in Figure 12–3. This rung will appear to work normally; that is, when the START is momentarily pressed, relay RUN switches ON and remains ON. When STOP is pressed, RUN switches OFF. However, consider what happens when both START and STOP are pressed simultaneously. For this program, START will override STOP, and RUN will switch on as long as START is pressed.

FIGURE 12–3
Unsafe Start/Stop Program

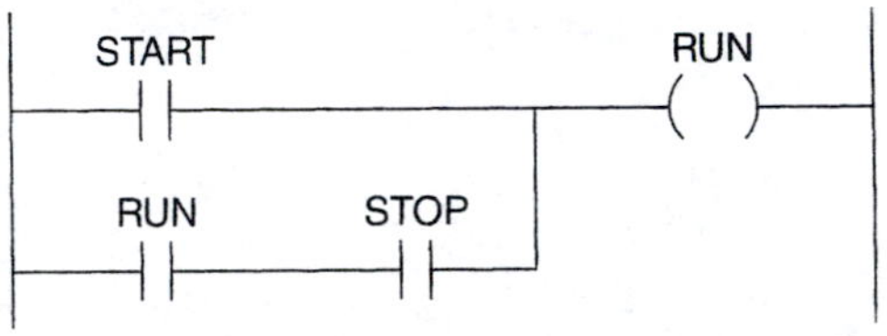

Now consider an improved version of this program shown in Figure 12–4. Notice that by moving the STOP contact into the main part of the rung, the START switch can no longer override the STOP. This program is considered safer than the one in Figure 12–3.

FIGURE 12–4
Improved Start/Stop Program with Overriding Stop

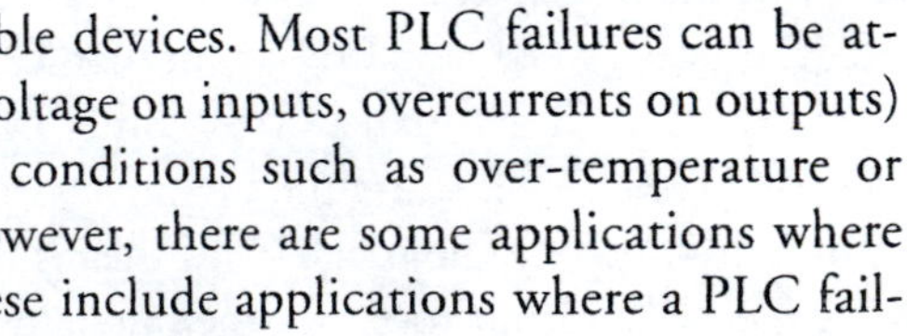

Generally, PLCs are extremely reliable devices. Most PLC failures can be attributed to application errors (overvoltage on inputs, overcurrents on outputs) or extremely harsh environmental conditions such as over-temperature or lightning strike, to name a few. However, there are some applications where even more reliability is desired. These include applications where a PLC failure could result in injury or loss of life. For these applications, the designer must be especially careful to consider what will happen if power fails and what will happen if the PLC should fail with one or more outputs stuck ON or stuck OFF.

Having a power failure on a PLC system is a situation that can be handled by fail-safe design. However, the situation in which a PLC fails to operate correctly can be catastrophic, and no amount of fail-safe design using a single PLC can prevent this. This situation can be best handled by using redundant PLC design. In this case, two identical PLCs are used that are running identical programs. The inputs of the PLCs are wired in parallel, and the outputs of the PLC are wired in series (to do this, the outputs must be of the mechanical relay type).

Consider the redundant PLC system shown in Figure 12–5. For this system, both PLCs are running the program shown in Figure 12–4. When PB1 is pressed, IN1 on both PLC1 and PLC2 is energized. Since both PLCs are running the same program, they will both switch ON their OUT1 relay. Since PLC1 OUT1 and PLC2 OUT1 are connected in series, when they both switch ON, relay CR1 will be energized. However, assume that the PLC1 OUT1 relay becomes stuck ON because of either a relay failure or a PLC1 firmware crash. In this case, assuming PLC2 is still operating normally, its OUT1 will also continue to operate normally switching CR1 ON and OFF properly. Conversely, if PLC1 OUT1 fails in the stuck OFF position, then CR1 will not operate and the machine will fail to run. In either case, failure of one PLC does not create an unsafe condition.

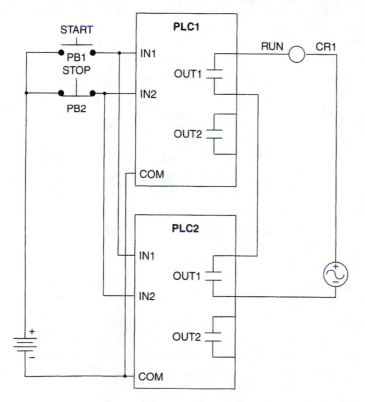

FIGURE 12–5

Increasing Reliability by Redundant PLC Design

One drawback to the redundant system shown in Figure 12–5 is that if one of the relays fails in the stuck ON condition, the system will continue to function normally. Although this is not hazardous, it defeats the purpose of having two PLCs in the system, which is to increase the reliability and safety. It would be helpful to have the PLCs identify when this occurs and give some indication of a PLC fault condition. This is commonly done by a method called **output readback.** In this case, the inputs and outputs must be of the same voltage type (e.g., 120 VAC), and additional inputs are purchased for the PLCs so that outputs can be wired back to unused inputs. Then, additional code is added to the programs to compare the external readback signal to the internal signal that produced the output. This is done using the disagreement (XOR) circuit. Any disagreement is cause for a fault condition.

12-4 Safety Interlocks

When designing control systems for heavy machinery, robotics, or high-voltage systems, it is imperative that personnel are prevented from coming in contact with the equipment while it is energized. However, since maintenance personnel must have access to the equipment from time to time, it is necessary to have doors, gates, and access panels on the equipment. Therefore, it is a requirement for control systems to monitor the access points to assure that they are not opened when the equipment is energized, or conversely, that the equipment cannot be energized while the access points are opened. This monitoring method is done with **interlocks.**

Interlock Switches

The simplest type of interlock uses a limit switch. For example, consider the door on a microwave oven. While the door is opened, the oven will not operate. Also, if the door is opened while the oven is operating, it will automatically switch off. This happens because a limit switch is positioned inside on the door latch such that when the door is unlatched, the switch is opened, and power is removed from the control circuitry. The most common drawback to **switch interlocks** is that they are easy to defeat. Generally, a match stick, screwdriver, or any other foreign object can be jammed into the switch mechanism to force the machine to operate. Designers go to great lengths to design interlock mechanisms that cannot be defeated.

More sophisticated interlock systems can be used in lieu of limit switches. These include proximity sensors, keylocks, optical sensors, magnetically operated switches, and various combinations of these. For example, consider the interlock shown in Figure 12–6. This is a combination deadbolt and interlock switch. When the deadbolt is closed, as shown in Figure 12–6a, a plunger and springs activate a switch that enables the machine to be operated. When the deadbolt is opened, as shown in Figure 12–6b, the switch opens also.

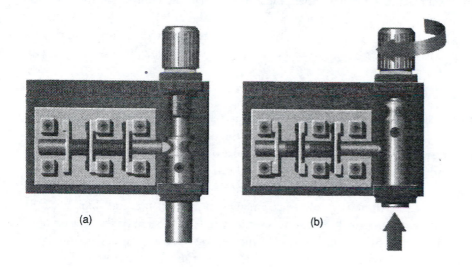

(a)   (b)

FIGURE 12–6
*Deadbolt Interlock
Switch
(Courtesy of
Scientific
Technologies, Inc.)*

Pressure Sensitive Mat Switch

The **pressure sensitive mat switch** is simply a mat that will close an electrical connection when force is applied to the mat. This is illustrated in Figure 12–7.

FIGURE 12–7
*Pressure Sensitive
Mat Switch
(Courtesy of
Scientific
Technologies, Inc.)*

They are available in many sizes and can be used for a wide variety of safety applications. For example, if used within a fenced area in which a robot operates, a mat can sense when someone has stepped within reach of the robot arm and can disable the robot. Another application is to place a mat in front of the operator panel of a machine. Then, connect the mat so that if the contacts in the mat are not closed, the machine will not run. Keep in mind that if they are used in "dead-man's switch" applications such as this, they can be defeated by simply setting a heavy object on the mat.

Pull Ropes

City transit buses use **pull ropes** on each side of the bus so that a rider can pull the rope, which switches on a buzzer notifying the driver that they wish to get off at the next stop. The technique is popular because only one switch is needed per pull rope, and the variable length of the rope can put it within reach over a large area. In short, it is a simple, reliable, and inexpensive mechanism. This same idea can be applied to machine controls as an emergency shut-off method. Figure 12–8 shows a pull rope interlock. The rope is securely fastened to a fixed thimble on one end and to a pull switch on the other. A turnbuckle tensioner allows for the slack to be adjusted out of the rope, and eye bolts support the rope. If necessary, the rope can even be routed around corners using pulleys. In use, the switch is N/C and connected into the controls as an additional STOP switch. When the rope is pulled by anyone, the switch contacts open, and the machine stops.

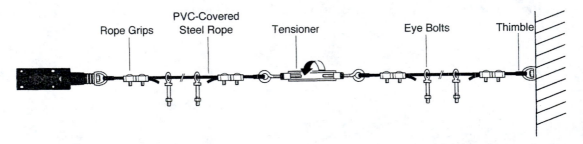

FIGURE 12–8
Pull Rope Interlock Switch
(Courtesy of Scientific Technologies, Inc.)

Light Curtains

When it is imperative that an operator's hands stay out of certain areas of a machine while it is operating, one excellent way to assure this is by using a **light curtain.** It is basically a "spinoff" of the thru-beam optical sensor; however, instead of the beam being a single straight line, the beam is extended (by scanning) to cover a plane, and the receiver senses the scanning beam over the entire plane. Any object intersecting the light curtain plane causes the electronic circuitry in the light curtain to output a discrete signal that can be used by the control circuitry to shut down the machine. The light used in this system is infrared and pulsed. Since it is infrared, it is invisible and does not distract the machine operator. Since the light is pulsed, it is relatively immune to fluorescent lights, arc welding, sunlight, and other light interference.

As shown in Figure 12–9, light curtains generally come in three parts—the transmitter wand, the receiver wand, and the power supply and logic (electronic) enclosure. The transmitter and receiver wands strictly produce and detect the infrared beams that make up the plane of detection. They are connected by electrical cables to the electronic enclosure, which contains all the necessary circuitry to operate the wands, power the system, make logical decisions based on the interrupted beams, and provide relay outputs. This is a complete "turn-key" stand-alone system. The designer simply supplies ac line power and connects the relay outputs to the machine controls.

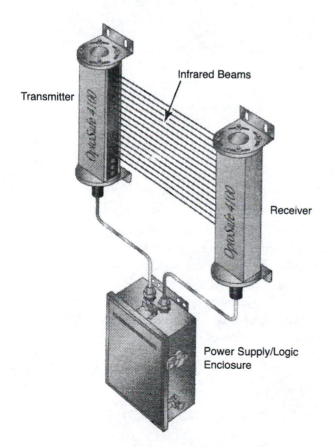

Transmitter

Infrared Beams

Receiver

Power Supply/Logic
Enclosure

FIGURE 12–9

*Light Curtain
Components
(Courtesy of
Scientific
Technologies, Inc.)*

If interlock coverage is needed over several planes, the transmitter and receiver wands of the light curtain can be cascaded as shown in Figure 12–10. In this case, two wands are cascaded on each end of the work area: one pair of transmitter wands and one pair of receiver wands. One pair of vertically positioned wands provides a vertical light curtain directly in front of the operator, and a second pair of horizontally positioned wands provides a horizontal light curtain underneath the work area. By doing this, the designer can realize some cost savings, since all the wands can be operated by one electronics enclosure. The logic circuitry in the electronics enclosure of the light curtain allows for an enable/disable discrete input so that the light curtain can be temporarily disabled to allow for the operator to load and unload the machine without tripping the curtain alarm.

FIGURE 12–10

Cascaded Light Curtains (Courtesy of Scientific Technologies, Inc.)

Summary

As we have seen, system integrity and system safety are important considerations for the system designer, not only when designing the system but also when specifying the types of equipment and enclosures to purchase. Environmental conditions such as weather, splashing water, corrosive agents, submersion, and human intrusion by tools and fingers can cause damage both to the equipment and the operators and maintenance personnel, and must be considered by the designer. If the system is performing a crucial task, fail-safe design must be considered. Finally, if the equipment being controlled has the capability to harm personnel, steps must be taken using interlocking schemes to assure that personnel cannot be in dangerous areas while the machine is in operation.

Review Questions

1. An electrical equipment enclosure is to be mounted on an above-ground swimming pool pump. It will be exposed to outdoor weather and an occasional splash of chlorinated (corrosive) pool water. Obviously, we would need to keep fingers and tools from fitting through any openings in the box. (a) What is the optimum NEMA number for this application? (b) What is the optimum IEC IP number for this application?

2. Select the best IEC IP number for an electrical box used in a textile mill. It must be dust tight and be able to withstand an occasional water hosedown by the cleaning crew.

3. Convert the IEC rating IP64 to the nearest equivalent NEMA rating.

4. Redraw the ladder diagram of Figure 12–4 to allow for the addition of a N/C pull rope switch that will disable the RUN output when it is pulled.

5. Devise a ladder diagram to control a production line conveyor. There will be a momentary N/O START pushbutton connected to IN1, a momentary N/C STOP pushbutton connected to IN2, and a N/C PULL ROPE switch connected to IN3. OUT1 will operate the conveyor MOTOR, and OUT2 will operate a warning HORN. When START is pressed, the HORN will sound for 10 seconds. After 10 seconds, the HORN will stop sounding, and the MOTOR will switch ON. Once started, the MOTOR can be stopped by pressing the STOP pushbutton or pulling on the pull rope. The STOP and PULL ROPE must be overriding inputs.

6. A light curtain is connected to a PLC such that when the light curtain is interrupted, IN1 of the PLC is turned OFF. OUT1 of the PLC is connected to a horn such that when OUT1 is ON, the horn sounds. Draw the ladder rung of Figure 12–4, and then add additional code so that if RUN is ON and the light curtain is interrupted, RUN will switch OFF, and the horn will pulse repeatedly ON for 0.5 second and OFF for 0.5 second. The horn will continue to sound, and RUN cannot be reactivated until the system is reset by pressing the STOP switch.

Logic Symbols

LOGIC SYMBOLS

LOGIC ELEMENT	AND	OR	NOT	NAND	NOR
LOGIC ELEMENT FUNCTION	OUTPUT IF ALL CONTROL INPUT SIGNALS ARE ON	OUTPUT IF ANY ONE OF THE CONTROL INPUTS IS ON	OUTPUT IF SINGLE CONTROL INPUT SIGNAL IS OFF	OUTPUT IF ALL CONTROL INPUT SIGNALS ARE ON	OUTPUT IF ANY OF THE CONTROL INPUTS ARE ON
MIL-STD-806B AND ELECTRONIC LOGIC SYMBOL					
ELECTRICAL RELAY LOGIC SYMBOL					
ELECTRICAL SWITCH LOGIC SYMBOL					
ASA (JIC) VALVING SYMBOL					
ARO PNEUMATIC LOGIC SYMBOL					
NFPA STANDARD					
BOOLEAN ALGEBRA SYMBOL	$(\ )\cdot(\ )$	$(\ )+(\ )$	$\overline{(\ )}$	$\overline{(\ )\cdot(\ )}$	$\overline{(\ )+(\ )}$
FLUIDIC DEVICE TURBULENCE AMPLIFIER					

(Courtesy of American Technical Publishers, Inc.)

Industrial Electrical Symbols

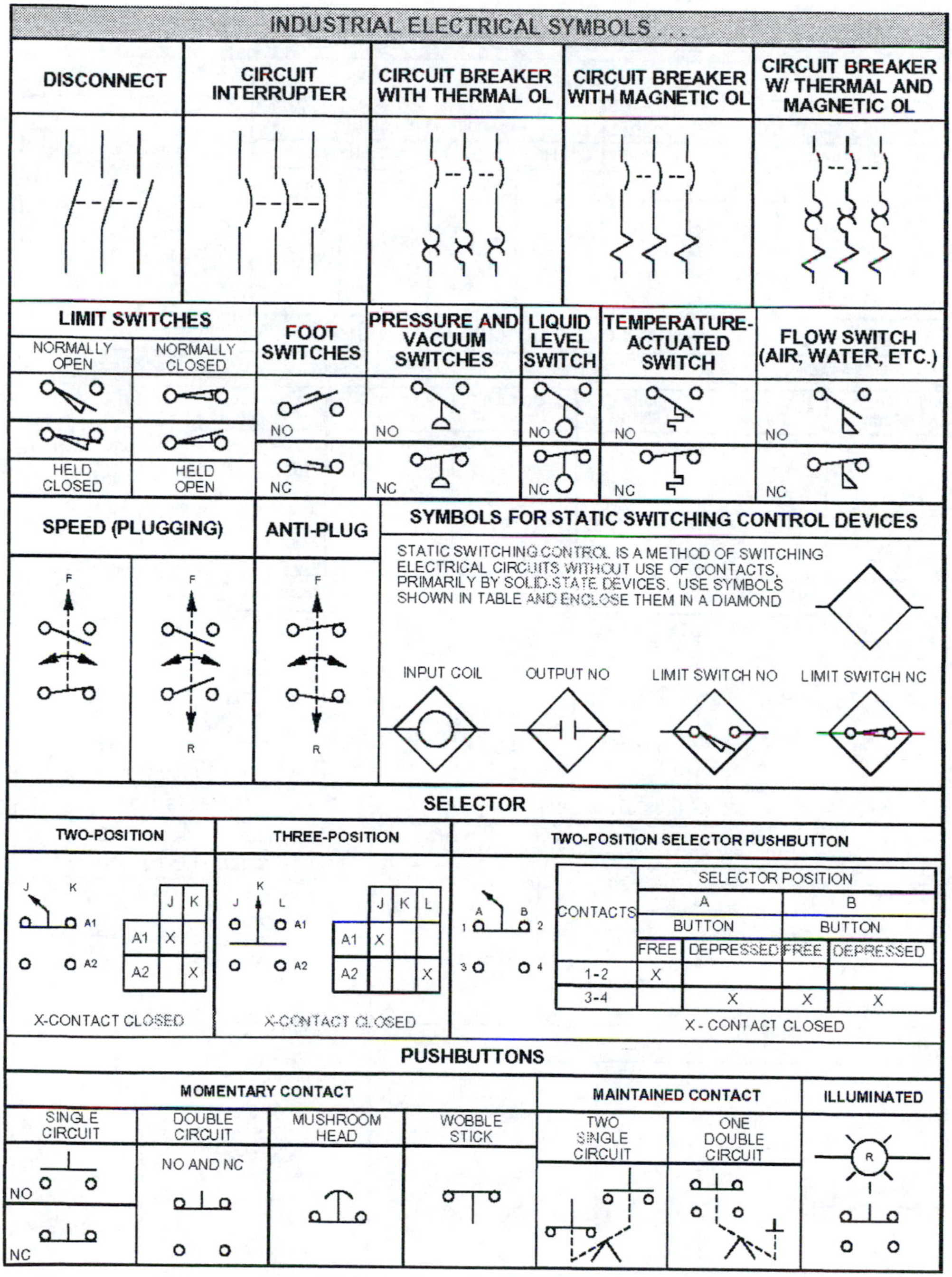

INDUSTRIAL ELECTRICAL SYMBOLS . . .

DISCONNECT	CIRCUIT INTERRUPTER	CIRCUIT BREAKER WITH THERMAL OL	CIRCUIT BREAKER WITH MAGNETIC OL	CIRCUIT BREAKER W/ THERMAL AND MAGNETIC OL

LIMIT SWITCHES

NORMALLY OPEN	NORMALLY CLOSED	FOOT SWITCHES	PRESSURE AND VACUUM SWITCHES	LIQUID LEVEL SWITCH	TEMPERATURE-ACTUATED SWITCH	FLOW SWITCH (AIR, WATER, ETC.)
HELD CLOSED	HELD OPEN	NO / NC	NO / NC	NO / NC	NO / NC	NO / NC

SPEED (PLUGGING) | ANTI-PLUG | SYMBOLS FOR STATIC SWITCHING CONTROL DEVICES

STATIC SWITCHING CONTROL IS A METHOD OF SWITCHING ELECTRICAL CIRCUITS WITHOUT USE OF CONTACTS, PRIMARILY BY SOLID-STATE DEVICES. USE SYMBOLS SHOWN IN TABLE AND ENCLOSE THEM IN A DIAMOND

INPUT COIL   OUTPUT NO   LIMIT SWITCH NO   LIMIT SWITCH NC

SELECTOR

TWO-POSITION	THREE-POSITION	TWO-POSITION SELECTOR PUSHBUTTON

X-CONTACT CLOSED

CONTACTS	A BUTTON FREE	A BUTTON DEPRESSED	B BUTTON FREE	B BUTTON DEPRESSED
1-2	X			
3-4		X	X	X

X - CONTACT CLOSED

PUSHBUTTONS

MOMENTARY CONTACT				MAINTAINED CONTACT		ILLUMINATED
SINGLE CIRCUIT	DOUBLE CIRCUIT NO AND NC	MUSHROOM HEAD	WOBBLE STICK	TWO SINGLE CIRCUIT	ONE DOUBLE CIRCUIT	

NO / NC

(Courtesy of American Technical Publishers, Inc.)

...INDUSTRIAL ELECTRICAL SYMBOLS...

CONTACTS

INSTANT OPERATING				TIMED CONTACTS - CONTACT ACTION RETARDED AFTER COIL IS:			
WITH BLOWOUT		WITHOUT BLOWOUT		ENERGIZED		DE-ENERGIZED	
NO	NC	NO	NC	NOTC	NCTO	NOTO	NCTC

OVERLOAD RELAYS

THERMAL	MAGNETIC

SUPPLEMENTARY CONTACT SYMBOLS

SPST NO		SPST NC		SPDT		TERMS
SINGLE BREAK	DOUBLE BREAK	SINGLE BREAK	DOUBLE BREAK	SINGLE BREAK	DOUBLE BREAK	SPST SINGLE-POLE, SINGLE-THROW
DPST, 2NO		DPST, 2NC		DPDT		SPDT SINGLE-POLE, DOUBLE-THROW
SINGLE BREAK	DOUBLE BREAK	SINGLE BREAK	DOUBLE BREAK	SINGLE BREAK	DOUBLE BREAK	DPST DOUBLE-POLE, SINGLE-THROW

TERMS:
- SPST SINGLE-POLE, SINGLE-THROW
- SPDT SINGLE-POLE, DOUBLE-THROW
- DPST DOUBLE-POLE, SINGLE-THROW
- DPDT DOUBLE-POLE, DOUBLE-THROW
- NO NORMALLY OPEN
- NC NORMALLY CLOSED

METER (INSTRUMENT)

INDICATE TYPE BY LETTER	TO INDICATE FUNCTION OF METER OR INSTRUMENT, PLACE SPECIFIED LETTER OR LETTERS WITHIN SYMBOL.			
	AM or A	AMMETER	VA	VOLTMETER
	AH	AMPERE HOUR	VAR	VARMETER
	µA	MICROAMMETER	VARH	VARHOUR METER
	mA	MILLAMMETER	W	WATTMETER
	PF	POWER FACTOR	WH	WATTHOUR METER
	V	VOLTMETER		

PILOT LIGHTS

INDICATE COLOR BY LETTER	
NON PUSH-TO-TEST	PUSH-TO-TEST

INDUCTORS

IRON CORE
AIR CORE

COILS

DUAL-VOLTAGE MAGNET COILS		BLOWOUT COIL
HIGH-VOLTAGE	LOW-VOLTAGE	
LINK	LINKS	

(Courtesy of American Technical Publishers, Inc.)

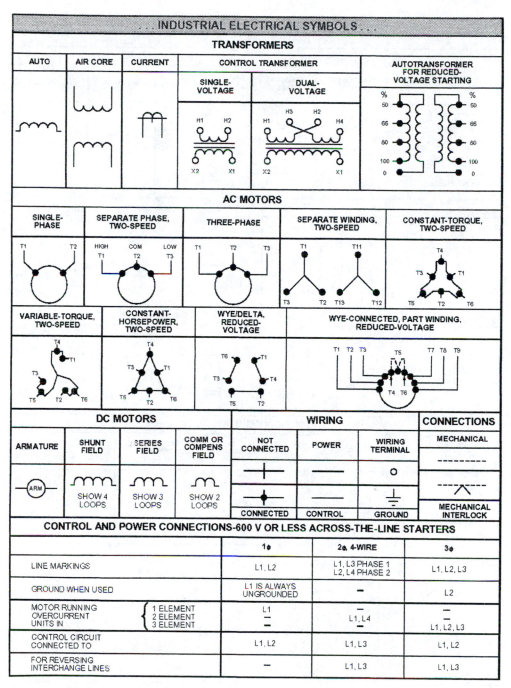

... INDUSTRIAL ELECTRICAL SYMBOLS ...

TRANSFORMERS

AUTO	AIR CORE	CURRENT	CONTROL TRANSFORMER		AUTOTRANSFORMER FOR REDUCED-VOLTAGE STARTING
			SINGLE-VOLTAGE	DUAL-VOLTAGE	

AC MOTORS

SINGLE-PHASE	SEPARATE PHASE, TWO-SPEED	THREE-PHASE	SEPARATE WINDING, TWO-SPEED	CONSTANT-TORQUE, TWO-SPEED

VARIABLE-TORQUE, TWO-SPEED	CONSTANT-HORSEPOWER, TWO-SPEED	WYE/DELTA, REDUCED-VOLTAGE	WYE-CONNECTED, PART WINDING, REDUCED-VOLTAGE

DC MOTORS / WIRING / CONNECTIONS

ARMATURE	SHUNT FIELD	SERIES FIELD	COMM OR COMPENS FIELD	NOT CONNECTED	POWER	WIRING TERMINAL	MECHANICAL
	SHOW 4 LOOPS	SHOW 3 LOOPS	SHOW 2 LOOPS	CONNECTED	CONTROL	GROUND	MECHANICAL INTERLOCK

CONTROL AND POWER CONNECTIONS—600 V OR LESS ACROSS-THE-LINE STARTERS

	1φ	2φ, 4-WIRE	3φ
LINE MARKINGS	L1, L2	L1, L3 PHASE 1 L2, L4 PHASE 2	L1, L2, L3
GROUND WHEN USED	L1 IS ALWAYS UNGROUNDED	—	L2
MOTOR RUNNING OVERCURRENT UNITS IN { 1 ELEMENT / 2 ELEMENT / 3 ELEMENT	L1 — —	— L1, L4 —	— — L1, L2, L3
CONTROL CIRCUIT CONNECTED TO	L1, L2	L1, L3	L1, L2
FOR REVERSING INTERCHANGE LINES	—	L1, L3	L1, L3

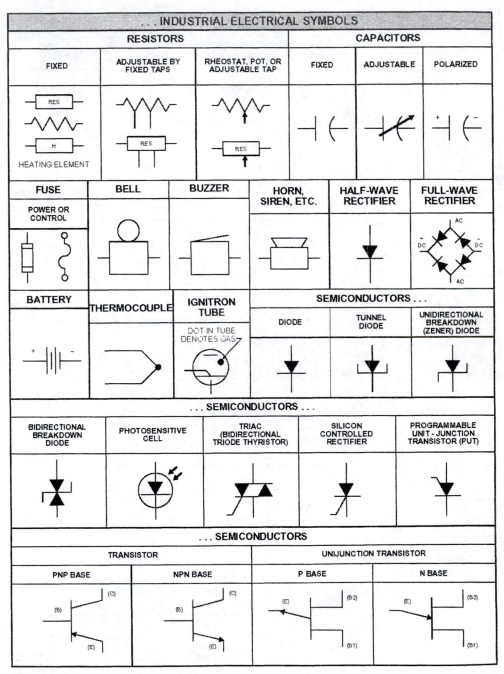

(Courtesy of American Technical Publishers, Inc.)

Bibliography

Allen Bradley, *SLC 500 Reference Manual*. Publ. 1747-6.15, 1996.

Hubert, Charles I. *Electric Machines: Theory, Application, Adjustment, and Control*, 2nd ed. Upper Saddle River, NJ: Prentice-Hall, 2002.

Omega Engineering, Inc. *Omega Complete Pressure, Strain, and Force Measurement Handbook and Encyclopedia*. Z-Technical Reference Section, 2001.

Omega Engineering, Inc. *Omega Complete Temperature Measurement Handbook and Encyclopedia*. Z-Technical Reference Section, 2001.

Omega Engineering, Inc. *Omega Flow and Level Measurement Handbook and Encyclopedia*. Z-Technical Reference Section, 2001.

Scientific Technologies, Inc. *Engineering Guide to Machine and Process Safeguarding*. 2000.

Stanley, William D. *Operational Amplifiers with Linear Integrated Circuits*, 4th ed. Upper Saddle River, NJ: Prentice-Hall, 2001.

Stenerson, Jon. *Fundamental of Programmable Logic Controllers, Sensors, & Communications*. Upper Saddle River, NJ: Prentice-Hall, 1994.

Webb, John W., & Reis, Ronald A. *Programmable Logic Controllers: Principles and Applications*, 4th ed. Upper Saddle River, NJ: Prentice-Hall, 1999.

Westinghouse Electric Company. *PC700/900 Programmable Controllers User's Manual*, 1984. Westinghouse.

Glossary

A

Absolute encoder   A device that measures angular or linear position and outputs it as a binary number. Absolute encoders do not need to be homed.

Absolute position   In rotary or linear encoders, the angular or linear position with respect to a known reference or home position.

Accelerometers   A transducer used to measure acceleration.

Actual   In a timer, the present time value. In a counter or sequencer, the present count value.

Analog sensor   A sensor that outputs a voltage or current that is proportional to the input parameter being sensed.

AND   A logical operation combining two or more binary values. The result (output) of the AND operation will be ON only when all of the input values are ON. In PLC mnemonic programming, the command to AND the next entry with the contents of the stack.

ANDI or AND NOT   In PLC mnemonic programming, the command to invert the next entry and then AND it with the contents of the stack.

Anti-repeat   A machine controls design technique that prevents a machine from performing more than one cycle each time the RUN switch is pressed.

Anti-tie down   Used in conjunction with two-handed run systems. A design technique that prevents a machine from operating if the operator has tied down or taped down one of the two switches in an attempt to operate the machine with one hand.

Autotune   In PID systems, a mode in which the system automatically determines the tuning constants required to achieve a stable system.

Auxiliary contact   A relay contact on a motor starter that is used for control logic purposes. Auxiliary contacts are rated at a lower current than the main contacts.

B

Backing or carrier   In a strain gage, the flexible substrate upon which the strain gage is bonded.

Bang-bang sensor   A sensor that has one digital output signal. It provides an ON-OFF signal only.

Batch unit high limit or Batch unit preload   In a PID control system, limits that are placed on the CV (control variable) output to reduce the effects of integral windup. When the batch unit high limit or the batch unit preload values are reached, the PID discontinues calculating the integral function.

Beam angle   In ultrasonic and optical sensors, the conical shaped illuminated area in front of the sensor head that constitutes the sensing area. Beam angle is the angle formed by the two opposite sides of the area.

Bi-directional counter   A counter that is capable of both incrementing (counting up) and decrementing (counting down).

Bi-directional sequencer   A sequencer that will cycle both forward (count up) and backward (count down).

Bit resolution   In D/A converters, the number of parallel input bits. In A/D converters, the number of parallel output bits. Generally speaking, converters with larger bit resolutions are more precise.

Branch   A portion of an electrical controls diagram rung or a ladder logic rung that provides a parallel path.

Branch close   In some PLCs, a programming command to end a branch, usually abbreviated BND or BCL.

Branch open   In some PLCs, a programming command to begin a branch, usually abbreviated BRO.

Brush drop   In motors and generators, the voltage drop across the brush and the commutator or slip rings.

C

Calibration factor   In proportional sensors, the ratio of the output to the input. For example, a 10 volt, 100 psi pressure sensor has a calibration factor of 10 V/100 psi or 0.1 Volt/psi.

Capacitive proximity sensor   A proximity sensor that detects the target material as a component of a capacitor. There are two types of capacitive proximity sensors: the conductive and the dielectric. The dielectric capacitive proximity sensor is capable of sensing non-conductive targets.

Carrier or backing   In a strain gage, the flexible substrate onto which the strain gage is bonded.

Cascaded timers   Two or more timers connected such that when the first times out, the second is started, when the second times out, the third is started, and so forth. The result is a set of signals that can be used to control a timed sequence of events that is initiated by energizing the first timer.

Closed-loop system   A control system in which the machine performance is monitored and fed back to the controller, allowing it to fine tune the system's performance.

Commutation diode   Also called a freewheeling diode. In pulse width modulated voltage controllers, a diode that provides a path for current flow during the time the switching device is off.

Conductive type capacitive proximity sensor   A capacitive proximity sensor that has one metal plate in the sensor head. It relies on the conductive target material acting as the other plate of a capacitor, with the capacitance varying with respect to the distance.

Contact logic   In PLC programming, the arrangement of N/O and N/C contacts on the left side of a rung. The PLC evaluates the contact logic in a rung to determine if the rung is ON or OFF.

Contactor   A heavy-duty relay.

Control transformer   A transformer used in control systems to reduce the line voltage to a value that is safe to use for powering control circuitry.

Control variable (CV)   In a closed loop control system, the signal that directly controls the machine.

Count line   In PLC programming, the input to a counter that causes the counter to advance by one when the line transitions from OFF to ON.

Counter   A PLC programming function capable of incrementing or decrementing its count value when given a transitional input. Counters are used to count numbers of input transitions and perform various operations at predetermined count values.

Cycle   In an industrial machine, one complete operation (e.g., stamp, form, cut, etc.). In PLC operation, one scan.

D

Dark-off   An optical proximity sensor that switches OFF when no light is striking the sensor head.

Dark-on   An optical proximity sensor that switches ON when no light is striking the sensor head.

Data available or data valid   In an absolute optical encoder, a strobe bit. A one-bit signal indicating that the parallel binary data is stable and can be sampled.

Deadband In an ultrasonic proximity sensor, the area near the sensor head in which a target cannot be detected. In any discrete sensor, the range between the ON and OFF switchpoints. For example, an oven thermostat that switches ON at 200° F and switches OFF at 195°F has a 5°F deadband.

Delay-on A time delay relay that actuates its contacts a preset time after its coil is energized.

Delay-off A time delay relay that keeps its contacts actuated a preset time after its coil is de-energized.

Derivative time constant Denoted by the variable T_d. In PID control systems, the derivative gain constant k_d.

D flip flop A flip flop with two inputs, data (D) and clock (CLK). There are two types of D flip flop, the level-triggered and the edge triggered. The level-triggered D flip flop continuously loads the value on the D input any time that the CLK input is ON. The edge-triggered D flip flop samples and loads the value on the D input whenever the CLK transitions from OFF to ON.

Dielectric type capacitive sensor A capacitive proximity sensor that has two metal plates in the sensor head. It relies on the target material acting as a varying dielectric, which varies the capacitance between the two plates.

Difference function Also called a discrete derivative or numerical derivative. A method of approximating the derivative of a signal by taking discrete samples of the signal at regular time intervals. The difference function is the change in the signal value divided by the time interval.

Diffuse reflective optical sensor An optical proximity sensor that detects light reflected from the target object.

Digital sensor A sensor that has one digital output signal. It provides an ON-OFF signal only.

Digital signal processing (DSP) Also called discrete signal processing. A method of performing analog-like functions on signals using a digital sampling and processing system.

Disagreement circuit A two-input circuit or network. The result (or output) will be ON when the inputs are in opposite logical states. An exclusive-OR or XOR circuit.

Discrete derivative Also called a difference function or numerical derivative. A method of approximating the derivative of a signal by taking discrete samples of the signal at regular time intervals. The discrete derivative is the change in the signal value divided by the time interval.

Discrete integral Also called a numerical integral. A method of approximating the integral of a signal by taking discrete samples of the signal at regular time intervals. The discrete integral is the accumulated sum of each sampled value of the signal times the time interval.

Discrete sensor A sensor that has one digital output signal. It provides an ON-OFF signal only.

Discrete signal processing (DSP) Also called digital signal processing. A method of performing analog-like functions on signals using a digital sampling and processing system.

Dissimilar metals Two different metals welded together to form a thermocouple.

Duty cycle The percentage of time a periodic signal is on. Mathematically, the on-time divided by the period.

Dwell time The length of time a machine, PLC, or other device pauses between operations.

E

Electrical engineering PID Also called a parallel PID. A PID control system in which the proportional, derivative, and integral gain functions are connected in parallel.

Emergency stop (E-stop) switch A maintained red mushroom head switch on an industrial machine that, when pressed, removes all power from the controls circuitry and thus immediately and completely shuts down the machine.

Encoder   A transducer that converts angular or linear displacement to a digital signal or binary value.

ENTER   In PLC mnemonic programming, the keystroke that terminates a rung entry.

Error   In a closed loop control system, the signal resulting from subtracting the process variable PV from the setpoint SP.

Error amplifier   In a simple closed loop control system, the proportional gain amplifier.

Eutectic alloy   An alloy that, when heated, passes from the solid to the liquid state without passing through a mushy or pasty state.

Eutectic metallic alloy overload   A eutectic alloy device used in a motor starter overload which, when heated by an overload current, opens the overload contacts.

Exclusive-OR   A two-input circuit or network. The result (or output) will be ON when the inputs are in opposite logical states. A disagreement circuit.

Extended pushbutton   A pushbutton switch on which the end of the actuator protrudes beyond the switch guard making it easy to press the switch.

F

Fail-safe   A machine controls design philosophy in which, if any critical component fails, the machine will be totally disabled.

Feedback   In a PID control system, the process variable PV. Feedback is used to monitor the machine's performance.

First scan   Also called power-up scan. In PLC programming, a built in firmware function providing a contact that is ON for the first program scan, and OFF for all others.

Flasher   A controlling device with an output that cycles ON and OFF at predetermined intervals.

The ON and OFF dwell times are not necessarily equal.

Flip-flop   In PLC programming, a network of internal relays connected to form a one-bit storage device.

Flush pushbutton   A pushbutton switch on which the end of the actuator is flush with the switch guard making it difficult to accidentally actuate the switch.

Form   A relay contact configuration. There are three forms: Form A is N/O, Form B is N/C, and Form C is both N/O and N/C with a common terminal (commonly called single-pole double-throw).

Freewheeling diode   Also called a commutation diode. In pulse width modulated voltage controllers, a diode that provides a path for current flow during the time the switching device is off.

Function   In PLC programming, a preprogrammed operation performed by the PLC firmware, thereby making programming simpler. Examples include first-scan relays, oscillators, analog I/O operations, and mathematical operations.

G

Gage factor   The sensitivity of a strain gage, k, defined as the percentage change in resistance with respect to the strain. Gage factor is a dimensionless quantity.

Gray code   A binary numbering system devised such that any number can be changed to the next higher or lower number by changing only one bit.

Ground loop   A wiring condition in which two electrical systems share a common ground path. In many cases, a voltage drop in the ground, caused by currents of one electrical system, introduces undesirable noise or errors into the other system.

Guarded actuator   A method of protecting a switch actuator to prevent accidental operation. Guarded actuators are usually recessed within the guard.

H

High-speed timer   In a PLC, a timer that advances in 0.01 second increments.

Holding contact   Also called a sealing contact, seal-in contact, or latching contact. A contact within a rung that, once the rung is energized, allows the rung logic to keep itself energized.

Holding register   In a PLC, a general-purpose memory location used to store digital values. Holding registers may be 8, 12, 16 or more binary bits in length, and are usually processed by the PLC as values instead of individual bits. They are most commonly used for storing analog I/O values and as accumulators when performing mathematical operations. The actual value for timers, counters, and sequencers are also stored in holding registers.

Home, or home position   A mechanical position in a rotary or linear mechanical positioning system, from which all other angular or linear positions are measured. On a rotary or linear encoder, a dedicated output that is on when the encoder is in the home position.

Homing   A mode of operation of a rotary or linear mechanical positioning system in which it slowly advances toward end-of-travel in search of a sensor signal that indicates it has reached home position. Once home position is reached, the system stops motion and zeroes its internal counters.

Hunting   In closed loop control systems, a performance characteristic in which the system overshoots the setpoint one or more times before finally settling at a steady value.

Hot   The ungrounded, or non-neutral side of the AC line. In controls circuitry, the ungrounded terminal of the secondary winding of the control transformer.

Hysteresis or deadband   In any discrete sensor, the range between the ON and OFF switchpoints. For example, an oven thermostat that switches ON at 200° F and switches OFF at 195° F has a 5° F hysteresis, or deadband.

I

Ideal PID   A PID control system in which the output of the proportional gain amplifier is connected to the inputs of the derivative and integral gain functions.

Illuminated switch   A switch in which the actuator is translucent, and a lamp is positioned behind the actuator (e.g., an elevator call button).

Impact pressure   Also called dynamic pressure. In a pitot tube, one of the two pressures used to calculate the speed of fluid or gas flow. The impact pressure is the pressure at the end point of the tube, and it varies with the velocity of the fluid or gas.

Inch   In machine controls, similar to JOG, but usually used for conditions which allow the operator to very slowly position the machine in order to improve the accuracy of the operation. For safety reasons, machine control systems are designed so that the machine cannot be cycled by inching.

Inclinometer   A sensor that measures the angle of inclination with respect to horizontal.

Incremental encoder   A device that converts angular or linear displacement to a train of digital pulses. Most have two outputs, phase A and phase B, that are phased at 90° with respect to each other. By identifying the leading phase, the direction of motion can be determined.

Inductive prox   Short for inductive proximity sensor.

Inductive proximity sensor   A proximity sensor that senses a target object by subjecting it to an alternating magnetic field and measuring the eddy current losses in the target.

Integral time constant T_i   In a PID control system, the inverse of the integral gain. Mathematically, $T_i = 1/k_i$.

Integral wind-up Also called reset windup. In PID control systems, an undesirable condition in which the integral collects a large error during the time a machine is paused, resulting in large unpredictable transient performance when the machine is enabled.

Interlock In machine controls, a device designed to disable a machine in the event of an unsafe condition, such as an open access panel, an operator who has left their station, or a person who has strayed into an unsafe area. In PLC programming, a zone. A section of a program that can be enabled or disabled under control of a programmed function, usually a relay contact.

Interlock ON and Interlock OFF In PLC programming, the endpoints of a control zone or interlock.

Internal relays In PLC programming, software relays that may be used within PLC ladder programs to perform general logical operations. An internal relay does not physically exist, but instead is represented by a digital bit within the PLC memory. Once created (or defined), an internal relay consists of one coil, and as many N/O and N/C contacts as are needed in the program.

I/O update In a PLC, the process in which the digital values stored in the output image register are transferred to the output terminals, and the digital values on the input terminals are transferred to the input image register. I/O update occurs at the beginning of each scan.

J

JK flip flop A flip flop with three inputs, J, K, and clock (CLK). When J is ON, K is OFF, and the clock transitions from OFF to ON, the flip flop will switch ON. When J is OFF, K is ON and the clock transitions from OFF to ON, the flip flop will switch OFF. When both J and K are ON and the clock transitions from OFF to ON, the flip flop will toggle states.

Jog A condition in which the machine can be "nudged" a small amount to allow for the accurate positioning of raw material while the operator is holding the material.

L

Ladder diagram In machine controls, the electrical controls diagram. In PLC programming, the program. So named because they resemble a ladder with two uprights and one or more rungs.

Latching contact A seal-in contact, sealing contact, or holding contact. A contact within a rung that, once the rung is energized, allows the rung logic to keep itself energized.

LD or STR In mnemonic PLC programming, a programming command that begins a new rung, or combines the most recent entry with the contents of the programming stack.

LDI or LD NOT In mnemonic PLC programming, a programming command that inverts the next entry and places it on the stack.

Light curtain A safety system in which a curtain of light separates a machine and the operator. Interrupting the curtain of light will disable the machine.

Lighted pushbutton An illuminated pushbutton switch.

Load In electrical controls, a device that transforms electrical power to some other form (e.g., heater, lamp, relay coil, motor, solenoid). In PLC programming, the coil in a rung. Loads are always the rightmost elements in the rungs.

Load cell A transducer that converts force or weight to a voltage or current. Most load cells use strain gages as the internal force sensor. Load cells can also be used as accelerometers if the mass is a known quantity.

Logic sensor A sensor that has one digital output signal. It provides an ON-OFF signal only.

M

Main contacts   In a motor starter, the heavy duty contacts that control power to the motor.

Mains   In ac power systems, the power line.

Maintained pushbutton   Also called a push-push switch. When pressed, it remains on until pressed again (e.g., the on/off switch of a desktop computer).

Majority circuit   A circuit or network with two or more inputs. The result (output) will be ON when more than half of the inputs are ON.

Master control relay (MCR)   A relay contact that is inserted in the hot rail of a controls circuit ladder wiring diagram. When an MCR contact is open, it disables all the rungs below it.

Maximum range   In a proximity sensor, the maximum distance at which a target object can be detected.

MCR ON and MCR OFF   In PLC programming, the commands that define the endpoints of an MCR.

Microstrain or μstrain   Strain times 10^{-6}. Since strain for most materials is a very small value, representing it as microstrain eliminates the need for the power of ten.

Model reference   In PID control systems, a sophisticated version in which a digital model of the system performance is input to the controller, and then the PID dynamically tunes the system to maintain the same performance as the model.

Modularized PLC   A PLC that can be purchased and assembled as component modules. It has the advantage in being able to purchase only the components needed, and can be expanded later.

Momentary pushbutton   A pushbutton that, when pressed, remains on until it is released (e.g., the reset button of a desktop computer).

Mounting rack   The frame onto which modules of a modularized PLC are installed. The mounting rack provides mechanical support for the modules and the electrical connections between them.

Mushroom head pushbutton   A pushbutton with a mushroom-shaped actuator. It is designed to make a switch easy to operate, such as in the case of an emergency stop switch.

N

Neutral   Sometimes also called low-side of the power line. The neutral is connected (bonded) to earth ground at its source.

Non-retentive timer   A timer that automatically resets its actual value when its coil is de-energized.

Normal/maintenance selector   In machine controls, a selector switch that allows the machine to be placed in one of two operational modes: normal (for usual operation) and maintenance (for repairs, adjustments, and troubleshooting). Normal/maintenance selector switches usually have keyed actuators to prevent the selection of the maintenance mode by unauthorized personnel.

Normally closed contacts   The contacts of a relay that are closed when the relay coil is de-energized, and open when the coil is energized. Also called Form B contacts.

Normally open contacts   The contacts of a relay that are open when the relay coil is de-energized, and closed when the coil is energized. Also called Form A contacts.

NOT   In PLC mnemonic programming, a command that instructs the PLC to invert the next element.

NPN   In PLC hardware, a sinking output. A connection method in which current is flowing into the collector of the NPN output transistor when it is switched ON.

Numerical derivative   Also called a discrete derivative or difference function. A method of approximating the derivative of a signal by taking discrete samples of the signal at regular time intervals. The numerical derivative is the change in the signal value divided by the time interval.

Numerical integral   Also called a discrete integral. A method of approximating the integral of a signal by taking discrete samples of the signal at regular time intervals. The discrete integral is the accumulated sum of each sampled value of the signal times the time interval.

O

OFF   In machine controls and PLC programming, a logical state in which the coil of a relay is de-energized. When a relay is OFF, its N/O contacts are open, and its N/C contacts are closed.

Offset   Also called the error. In closed loop control systems, the difference between the steady-state machine operating point and the desired operating point. Also, the setpoint SP minus the process variable PV.

Offset binary   In bipolar A/D and D/A converters, a scaling method in which binary zero represents the lowest (most negative) analog voltage or current, and full scale (all ones) represents the most positive voltage or current.

ON   In machine controls and PLC programming, a logical state in which the coil of a relay is energized. When a relay is ON, its N/O contacts are closed, and its N/C contacts are open.

One-shot   In PLC programming, a relay that is connected so that it will be ON for only one scan each time it is triggered.

Open frame PLC   A PLC consisting of a single circuit board that is not housed in a case or box.

OR   A logical operation combining two or more binary values. The result (output) of the OR operation will be ON when any one or more input values are ON. In PLC mnemonic programming, the command to OR the next entry with the contents of the stack.

OR BLK   In PLC mnemonic programming, the command to OR the active area with the contents of the stack.

OR LD   In PLC mnemonic programming, the command to OR the active area with the contents of the stack.

ORI or OR NOT   In PLC mnemonic programming, the command to invert the next entry and then OR it with the contents of the stack.

Oscillation   In closed loop control systems, an undesirable condition in which the system continuously alternates above and below the setpoint.

Oscillator   In PLC programming, a relay that is connected so that it will be alternately ON for one scan, and OFF for one scan.

Output (CV) max and Output (CV) min   In a PID control system, limits that are placed on the CV (control variable) output to reduce the effects of integral windup. When the CV max or CV min values are reached, the PID discontinues calculating the integral function.

Output readback   In PLC systems, a method to increase reliability of a system by connecting outputs to inputs in order that the system can monitor its outputs for failure.

Overload   In a motor starter, the device that monitors the motor current and heats the overload spindle (which opens the overload contacts in the event of an overcurrent condition).

Overload contact   In a motor starter, the contacts that are controlled by the overloads.

Overload spindle   In a motor starter, the device that is heated by the overload heater and controls the overload contacts.

Overriding reset   In a controls system, a reset that takes control over all other switches. For example, in a system with separate ON and OFF pushbutton switches, pressing the OFF switch will not only power-down the machine, but it will also disable the ON switch. In this case, the OFF switch is the overriding reset.

P

Paddle wheel flow sensor   A fluid flow sensor that measures fluid velocity by the rotating speed of a paddle wheel inserted into the flow.

Palm head pushbutton, or palm switch   Same as a mushroom head pushbutton. A pushbutton with a mushroom-shaped actuator, generally with a smooth finish. It is designed for ease of use in cases where the switch must be operated many times each day, and can easily be actuated with the palm of the hand.

Parallel PID   Also called an electrical engineering PID. A PID control system in which the proportional, derivative, and integral gain functions are connected in parallel.

Physical components   Components in a control system that can be physically touched, as opposed to program components that exist within a PLC program.

Pitot tube   An instrument for measuring fluid or gas flow rate by measuring the impact pressure with respect to the static pressure. The pitot tube is most commonly used as the sensor in airspeed indicators for aircraft.

PNP   In PLC hardware, a sourcing output. A connection method in which current is flowing out of the collector of the PNP output transistor when it is switched ON.

Power rails   Same as rails, or uprights. The vertical conductors in an electrical controls diagram, with the hot rail on the left, and the ground rail on the right.

Power-up scan   Also called first scan. In PLC programming, a built-in firmware function providing a contact that is on for the first program scan, and off for all others.

Preset   In a timer, the predetermined time units that the timer will run. In a counter, the maximum count at which it will cease to count. In a sequencer, the sequence number at which it will recycle back to zero.

Pressure hammer   In fluid systems, a pressure transient caused by abruptly closing a valve in the system. The transient is caused by the fluid inertia. Pressure hammer is potentially destructive to system components, especially pressure transducers.

Pressure hammer snubber   In fluid systems, a small orifice that is inserted in a pipe immediately before a pressure sensor to prevent damage to the sensor caused by a pressure hammer.

Pressure sensitive mat switch   A safety device that closes its switch contacts when weight is applied to the mat.

Pressure transducer   A sensor that converts fluid or gas pressure to a proportional voltage or current.

Process variable   Also called feedback. In closed loop control systems, a signal that is proportional to and indicates the machine's performance.

Processor   The portion of a modularized PLC that contains the microprocessor, memory, programmer interface, and in some cases, the power supply.

Program components   Components that exist within a PLC program, as opposed to physical components in a control system that can be physically touched.

Programming unit   A hand-held keypad unit used to program and troubleshoot a PLC. Programming units are usually used in the field and preclude the need for a computer.

Proportional gain amplifier   In a PID control system, a linear amplifier that amplifies the error signal.

Proportional gain k_p   In a PID control system, the gain or amplification factor of the proportional gain amplifier.

Proportional sensor   A sensor that outputs a voltage or current that is proportional to the input parameter being sensed.

Pull rope   A machine safety device in which a rope is routed to all areas where personnel are located. The rope is connected so that anyone pulling the rope will open a safety switch and disable the machine.

Pulse-width modulator motor speed control A motor speed controller in which the average output voltage is varied by varying the duty cycle (pulse width) of the controller.

R

Rails In machine controls diagrams, the hot and ground vertical conductors. In PLC programming, the two uprights in a ladder program.

Rate damping Also called throttle damping. In machine control systems, a characteristic that limits the maximum speed at which the machine's performance can be changed. Rate damping is usually provided by the derivative function in a PID controller.

Reference designator A unique combination of letters and numbers that identify a particular component in a control circuit.

Relative position In an incremental position encoder (rotary or linear), the angular or linear displacement from the most recent known position.

Reset In PID control systems, the integral function.

Reset line In PLC programming, the input terminal on a retentive timer, counter, or sequencer that resets the device. Non-retentive timers do not have reset lines.

Reset time constant In PID control systems, the integral time constant T_i.

Reset wind-up Also called integral windup. In PID control systems, an undesirable condition in which the integral collects a large error during the time a machine is paused, resulting in large unpredictable transient performance when the machine is enabled.

Resolution In A/D and D/A converters, the voltage or current represented by one least-significant binary bit. In rotary and linear encoders, the smallest angle or linear distance that the device can detect.

Retentive timer A timer that retains the actual value when its time input is OFF. If re-energized, it continues timing from the most recent actual value. A retentive timer does not reset when stopped. It can only be reset by activating its reset line.

Retro-reflective optical sensor An optical proximity sensor that senses when a normally reflected beam of light is interrupted by a target object. Retro-reflective optical sensors require the use of a special polarizing reflector.

Reversible sequencer A sequencer that can cycle both forward (count up) and backward (count down). Reversible sequencers have either an UP/DOWN input and a COUNT input, or separate count UP and count DOWN inputs.

R-S flip flop or R-S latch A flip flop with set (S) and reset (R) inputs and a Q output. It is used as a simple one-bit storage device.

Run In machine controls, the state in which the machine is both ON and is CYCLING. In PLCs, the mode of operation when the PLC is executing the program and performing I/O updates.

Rung In machine controls circuits, a horizontal circuit connected between the hot and ground rails. In PLC programming, one horizontal row of elements between the two uprights of a ladder logic program.

S

Saturation In PID control systems, a setting which limits the integral windup.

Scan One program cycle of a PLC that includes I/O update and one execution of the ladder logic program.

Scan rate The rate at which a PLC is performing program execution and I/O updates, usually measured in milliseconds per kilobits of program.

Scan time   The time for a PLC to perform one program scan and one I/O update, measured in milliseconds.

Seal-in contact, or sealing contact   Same as a holding contact or latching contact.

Sensor   A device used to detect or measure a physical property, such as temperature, distance, fluid flow, pressure, and so on.

Sequencer   A device or programming function that performs similar to a counter, but will count circularly. That is, if it is incrementing and reaches its preset value, it resets to zero and continues incrementing. If it is decrementing and reaches zero, it loads the preset value into the actual register and continues decrementing. Sequencers are used to control multiple operations that must be executed in a predetermined sequence.

Setpoint   In control systems, the desired operating point of the system. The setpoint is usually entered by the operator.

Shadow mask   In a rotary or linear encoder, a mask that overlays the etched glass that allows light to pass through when graduations on the glass are aligned with graduations on the shadow mask.

Shoe box PLC   A complete PLC that is contained in a single case, roughly resembling a shoe box. Generally they are less expensive and less powerful than modular PLCs.

Shrouded actuator   In a pushbutton, an actuator that is completely enclosed by a metal or plastic sleeve (called a shroud) which covers and protects the actuator and prevents accidental operation.

Single board PLCs   A PLC in which all electronic components are located on a single printed circuit board.

Single cycle   In machine controls, a mode of operation in which the machine performs one and only one operation each time the RUN switch is actuated.

Single-phasing condition   In three-phase motor systems, an undesirable condition in which one of the three phases is disconnected, and the motor continues to operate on the remaining two phases. The result is a decrease in available torque, low efficiency, overheating, and possible stalling and motor failure.

Single point grounding   A method of avoiding ground loop errors and interference by grounding all systems at a common tie point.

Sinking   A connection method in which current is flowing into the device when it is switched ON. A sinking input will be ON when current is flowing into the input terminal, and a sinking output that is ON will have current flowing into the output terminal.

Slidewire potentiometer   A wirewound linear potentiometer usually used as a position sensor for chart recorders and moving indicators.

Sourcing   A connection method in which current is flowing out of the device when it is switched ON. A sourcing input will be ON when current is flowing out of the input terminal, and a sourcing output that is ON will have current flowing out of the output terminal.

Stack   In PLC programming, a temporary storage area that retains the structure of the rung as it is being entered into the PLC, analogous to an accumulator in mathematical operations. PLC programming stacks are last-in, first-out (LIFO) stacks.

Standard timer   In a PLC, a timer that advances in 0.1 second increments.

Standby   A machine state in which power is applied to a machine and it is ON, but it is not running (or cycling).

Static pressure   In a pitot tube, the pressure on the side of the tube that is not a function of the fluid or gas flow. In an aircraft airspeed indicator system, the atmospheric pressure.

Stiction   Starting friction.

STO   In PLC mnemonic programming, a programming command that indicates the end of a rung.

Strain   The deformation of a material due to stress.

Stress   A force or pressure applied to a material causing it to deform (or strain).

Strobe   In an absolute optical encoder, a one-bit signal indicating that the parallel binary data is stable and can be sampled.

Switch interlocks   Safety switches installed on a system so that they will disable the machine when actuated. For example, the door switch on a washing machine is a switch interlock.

T

T flip flop   A flip flop with two inputs, T and clock (CLK). When T is OFF and the clock transitions from OFF to ON, the flip flop will remain in its previous state (i.e., it will do nothing). When T is ON and the clock transitions from OFF to ON, the flip flop will toggle states.

Thermal dispersion flow switch   A flow switch that uses the thermal dispersion properties of fluid to measure flow rate. The device heats the fluid as it passes and then measures the temperature of the fluid farther downstream to determine fluid velocity.

Throttle damping   Also called rate damping. In machine control systems, a characteristic that limits the maximum speed at which the machine's performance can be changed. Rate damping is usually provided by the derivative function in a PID controller.

Thru-beam optical sensor   An optical sensor that detects a target object that interrupts a light beam between the emitter and detector of the sensor.

Tilt gage   A proportional sensor that outputs a voltage or current that is proportional to the inclination with respect to horizontal.

Tilt switch   A discrete sensor that switches its output ON when the inclination exceeds its setpoint.

Timer, or time delay relay   In machine controls and PLC programming, a relay that delays switching on (delay-on) after its coil is energized, or delays switching off (delay-off) after its coil is de-energized.

Transducer   A sensor that converts one physical property to another. For example, an ultrasonic transducer converts ultrasonic sound waves to electrical signals, or vice versa.

Transitional contact   A changing contact. A transitional contact can be a contact that is switching from open to closed, or closed to open.

Turbine flow sensor   A fluid flow sensor that uses a propeller-like device to sense fluid motion.

Two-handed palming operation   Same as the two-handed run, but using palm switches that are actuated by the palms of the hands.

Two-handed run   A machine controls design technique in which two switches must be pressed simultaneously in order to cause a machine to cycle. The purpose is to occupy both of the operator's hands to keep them away from the machine, thus preventing them from being injured.

U

UP/DN   In PLC programming, an input on a bidirectional counter or sequencer that controls the direction it will count or advance.

Uprights   In PLC programming, the vertical rails of a ladder program.

Usable sensing area   In an ultrasonic proximity sensor, a cone shaped area in front of the sensor head. A target object within the usable sensor area will be detected by the sensor.

V

Variable frequency drive   An ac induction motor controller system that varies both the frequency and rms voltage applied to the motor in order to control its speed.

Voltage resolution For an analog-to-digital converter, or digital-to-analog converter, the voltage change that is equal to one least-significant binary bit (LSB).

W

Wire marker A label attached onto the end of a wire that identifies the wire number or name.

Z

Zone A section of a machine controls circuit, or a portion of a PLC ladder program that can be enabled or disabled by a controlling function, usually a relay contact.

Zone begin In PLC programming, a marker inserted into a ladder logic program indicating the beginning of a control zone.

Zone end In PLC programming, a marker inserted into a ladder logic program indicating the end of a control zone.

Index

Absolute encoder, 200–204
Absolute position, 199
Acceleration sensors, 193–194
Accelerometers, 193
Accelerometer strain gage signal
    conditioning, 194
ac motor overload protection,
    245–247
ac motor starter, 243–244
ac power wiring, 117
ACTUAL, 83
A/D input, 140–144
Adjust and observe tuning method,
    231–233
Advanced programming techniques,
    72–102
    cascading timers, 90–91
    control zones, 96–98
    counters, 83–85
    flashers, 91–92
    flip flops, 73–82. *See also* Flip
      flops
    master control relays, 96–98
    order of program execution, 73
    sequences, 85–86
    timed one shot, 93–94
    timed sequencer, 94–96
    timers, 87–91
Always de-energized coil, 66
Always energized coil, 66
Always-on/always-off contacts,
    65–66
Analog (D/A) output, 144–145
Analog data handling, 145
Analog I/0, 139–144
    analog (A/D) input, 140–144
    analog (D/A) output, 144–145
    constant offset error, 146
    data handling, 145

    percentage offset error, 146
    unstable reading, 147
Analog input and output modules,
    40
Analog input units, 43
Analog output units, 43
Analog (proportional) sensors, 151,
    152
Analog-to-digital (A/D) input,
    140–144
AND ladder diagram, 51
AND ladder rung, 104–105
AND lamp circuit, 17
AND-OR-AND logic, 107
AND-OR circuit, 20
AND-OR lamp circuit, 20
Angle position sensors, 194–196
    absolute encoder, 200–204
    incremental encoder, 196–199
    slotted disk/opto-interrupter,
      194–196
Anti-repeat, 24
Anti-tie down, 24
ASCII relay symbols, 11
Automatic one shot, 74
Autotuning PID systems, 239–240
Auxiliary contact, 243
AWG14, 15
AWG16, 15

Backing, 177
Bang-bang sensors, 151
Basic control circuit, 14
Batch unit high limit, 227
Batch unit preload, 227
Beam angle, 162
Bellows switch, 181–182
Bibliography, 281
Bi-directional counter, 85

Bi-directional sequencers, 86
Bi-metallic switch, 170–172
Binary output encoders, 200–204
Bipolar converters, 142–144
Bit resolution, 141
Boolean logic, 16–22
    AND, 17–18
    OR, 18–19
    AND OR/OR AND, 19–21
Branch
    complex, 110–112
    defined, 19
    simple, 107–110
Branch close, 108
Branch open, 108
Brush drop, 250

C, 15
Calibration factor, 185
Cam-operated limit switch, 26
Cam switches, 8
Capacitive proximity sensors,
    159–161, 176
Carrier, 177
Cascaded light curtains, 272
Cascaded timers, 90
CB, 16
Central processing unit (CPU), 39,
    42
Circuit
    combined, 29
    disagreement, 59
    majority, 59–61
    single-cycle, 27
Closed-loop control, 212–241
    derivative function, 220–224
    hunting, 217
    integral function, 225–228
    offset, 215

oscillation, 217
PID control. *See* PID control
problems, 214–217
simple closed-loop systems, 212–214
Combined circuit, 29
Common relay wiring, 129
Commutation diode, 250
Complex branches, 110–112
Complex ladder rung, 110
Conductive capacitive proximity sensors, 160
Constant offset errors, 146
Contact logic, 63
Contactor, 9
Control transformer, 2–3
Control variable (CV), 213
Control zones, 30–31, 96–98
Converting gray to binary, etc., 293
Counter, 83–85
Count line, 84
CPU, 39, 42
CR, 15
CV, 213
CYCLE, 32

D/A output, 144–145
Dark-off, 164
Dark-on, 164
Data available, 202
Data valid, 202
dc motor armature model, 251
dc motor controller, 249–255
dc motor control with ac power source, 252–255
dc motor control with dc power source, 249–251
dc motor speed control, 251
dc power wiring, 118
Deadband, 162, 176
Deadbolt interlock switch, 269
Delay-off timer (TOF) relay, 13
Delay-on timer (TON) relay, 12
Derivative function, 220–224
Derivative time constant (T_d), 229, 231, 232
Detent pushbutton, 6

D flip flop, 77–79
Diaphragm strain gage, 183
Dielectric capacitive proximity sensors, 160
Difference function, 220
Diffuse reflective optical sensor, 165
Digital sensors, 151
Digital signal processing, 214
Digital-to-analog (D/A) output, 144–145
Disagreement circuit, 59
Discrete derivative, 220
Discrete input units, 43
Discrete integral, 226
Discrete I/O modules, 40
Discrete output pressure switch symbol, 182
Discrete output temperature switch symbols, 172
Discrete position sensors, 150–168
connecting discrete sensors to PLC inputs, 154–155
proximity sensors, 156–167. *See also* Discrete position sensors
sensor output classification, 151–153
Discrete sensors, 151
Discrete signal processing, 214
Drag disk flow switch, 186–188
DSP, 214
Duty cycle, 249
Dwell time, 90

Electrical engineering PID, 219
Electrically erasable programmable read-only memory (EEPROM), 42
Electrical symbols, industrial, 276–279
Emergency Stop (E-Stop) switch, 6
Encoder
absolute, 200–204
defined, 170
glass scale, 206
incremental, 196–199

Equipment temperature considerations, 263–264
Error, 213
Error amplifier, 213
E-Stop switch, 6
Eutectic alloy, 245
Eutectic metallic alloy overload, 245, 246
Executing a ladder diagram, 45–47
Extended pushbutton, 5, 6
Externally triggered one shot, 76

F, 16
Fail-safe wiring/programming, 264–268
Feedback, 213
Firing angle, 253
Firmware flasher, 91
First output update, 44
First scan, 76
5-input majority circuit, 61
Flasher, 91–92
Flip flops, 73–82
D, 77–79
J-K, 81–82
one shot, 74–77
R-S, 73–74
T, 79–81
Float level switch, 175–176
Float switch, 175
Flow, measuring, 186–191
Fluid flow, 186
Fluid Flow and Level Handbook, 186
Flush pushbutton, 5
Force, measuring, 177–181
Form C relays with common output, 128
Forms, 127
4-bit binary encoder output signals, 201
4-bit binary optical absolute encoder disk, 200
Freewheeling diode, 250
Full-wave rectifier dc motor supply, 252

Function, 91
Fundamental PLC programming,
  50–71
  always-on/always-off contacts,
    65–66
  disagreement circuit, 59
  example (lighting control), 56–58
  holding contact, 64–65
  internal relays, 58
  ladder diagrams with multiple
    rungs, 67–69
  majority circuit, 59–61
  oscillator, 62–64
  physical vs. program components,
    51–56
Fuse, 3

Gage factor, 178
Gain error, 146
Gated oscillator, 64
Glass scale encoders, 206
Gray code, 202
Ground loop, 146
Ground test, 22–23
Ground test circuit, 22
Guarded actuator, 5

Half-range value, 143
Heater, 245
High-speed counter module, 40
High-speed timers, 87
Holding contact, 64–65
Holding register, 83
Home, 199
Home position, 195
Homing, 195
Hot, 117
Hunting, 217
Hysteresis, 176

Ideal PID, 219
IEC, 261
IEC enclosure rating table, 263
IEC/NEMA ratings, 261–263
Illuminated switch, 23
Image register, 44
Incandescent sensors, 163

INCH, 32
Inclination sensors, 192
Inclinometers, 192
Incremental encoder, 196–199
Incremental encoder output
  waveforms, 198
Indicator lamps, 8–9
Inductive proximity sensors,
  156–159
Industrial electrical symbols,
  276–279
In-line flow meter with proximity
  switch, 188
Input and output (I/O) modules, 40
Input units, 42–43
Input wiring, 119–121
Integral function, 225–228
Integral time constant (T_i), 229,
  231, 233
Integral wind-up, 227
Integrated circuit temperature
  probes, 174
Interlock Off, 96
Interlock On, 96
Interlocks. See Safety interlocks
Interlock switches, 268–269
Internal coil, 67
Internal relays, 58
International Electrotechnical
  Commission (IEC), 261
I/O modules, 40
I/O Update, 44
IR LED sensors, 163
Isolated contact wiring, 131
Isolated inputs, 124–126
Isolated relay output, 130

J-K flip flop, 81–82
JOG, 32

K_p adjustment, 231, 232
k_u, 235

L, 16
Ladder diagram, 13, 45
Ladder diagram fundamentals, 1–34
  basic components/symbols, 2–13

basic diagram framework, 13–14
boolean/relay logic, 16–22
combined circuit, 29
control zones, 30–31
ground test, 22–23
latch, 23–24
machine control terminology,
  31–32
MCRs, 30
reference designators, 15–16
single cycle, 26–28
two-handed anti-tie down, anti
  repeat, 24–26
wiring, 14–15
Latch, 23–24
Latch circuit, 23, 24
Latching contact, 23
LD command, 104
Light curtains, 270–272
Lighted pushbutton, 6
Lighting control system (example),
  56–58
Limit switches, 7–8
Linear displacement, measuring,
  207–209
Linear variable differential
  transformer (LVDT), 205–206
Liquid level, measuring, 175–177
Liquid level float switch, 175
Load, 18
Load cell, 193
Logic sensors, 151
Logic symbols, 275
LS, 15
LVDT, 205–206

M, 16
Machine control terminology, 31–32
Magnetostrictive linear displacement
  sensor, 207–209
Main contacts, 243
Mains, 244
Maintained pushbutton, 4
Majority circuit, 59–61
Majority table, 60
Master control relays (MCRs), 30,
  96–98

Maximum range, 162
MCR, 30, 96–98
MCR Off, 96
MCR On, 96
Measurement, 169–211
    acceleration, 193–194
    angle position sensors, 194–204.
        See also Angle position sensors
    flow, 186–191
    force, 177–181
    inclination, 192
    linear displacement, 204–208
    liquid level, 175–177
    pressure/vacuum, 181–186
    temperature, 170–174
Mechanically operated limit switch, 8
Memory, 42
Mnemonic programming code,
        103–114
    AND-OR-AND logic, 107
    complex branches, 110–112
    AND ladder rung, 104–105
    normally closed contacts,
        105–106
    OR-AND-OR logic, 109
    OR ladder rung, 106–107
    simple branches, 107–110
Model reference, 240
Model reference controllers, 240
Modularized PLC, 39
Momentary pushbutton, 4
Motor controls, 242–259
    ac motor overload protection,
        245–247
    ac motor starter, 243–244
    dc motor controller, 249–255
    motor starter selection criteria,
        247–248
    variable speed (frequency) ac
        motor drive, 255–258
Motor overload protection, 245–247
Motor starter selection criteria,
        247–248
Mounting rack, 39–40
MTW, 15
Mushroom head pushbuttons, 6
Mushroom head switches, 7

National Electrical Manufacturers
        Association (NEMA), 261
N/C contact, 4, 10, 58
NEMA, 261
NEMA enclosure rating table, 262
NEMA/IEC ratings, 261–263
Neutral, 117
Nichols, N. B., 234
N/O contacts, 4, 10, 105–106
Noisy reading, 147
Non-isolated input wiring, 122
Non-retentive timers, 87
Non-timed sequencer, 85
Normally closed (N/C) contact, 4,
        10, 58, 105–106
Normally open (N/O) contact, 4, 10
NORMAL/MAINTENANCE
        selector, 29
Notation. *See* Symbols
NPN sensor load connection, 153
NPN (sinking) output, 152
NPN units, 131, 132
Numerical derivative, 220
Numerical integral, 226

OFF, 32
Offset, 215
Offset binary, 142
OL, 16
ON, 32
¼ wave decay method, 234
One shot, 74–77
One shot ladder diagram, 76
Open frame PLCs, 37, 38
Open-loop method (ZN tuning),
        236–239
Optical proximity sensors, 163–167
Opto-interrupter, 194–196
Opto-isolators, 119–121
OR-AND circuit, 20
OR-AND lamp circuit, 21
OR-AND-OR logic, 109
OR circuit, 54
OR ladder rung, 106–107
OR lamp circuit, 18
Oscillation, 217
Oscillator, 62–64

Oscillator method (ZN tuning),
        235–236
Output (CV) max, 227
Output (CV) min, 227
Output image register, 44
Output readback, 268
Output units, 43
Output wiring, 126–127
Overload, 245
Overload contact, 243
Overload spindle, 246
Overriding reset, 88

Paddle wheel flow sensor, 189, 190
Palm head pushbutton, 6
Palm switch pushbuttons, 6
Parallel PID, 219
PB, 15
Percentage offset error, 146
Physical components, 51
PID adjustments, 231
PID control, 218–219, 228–229.
        See also Closed-loop control
    ideal PID, 219
    parallel PID, 219
    tuning the PID, 229–240. *See also*
        Tuning the PID
Pitot tube, 191
Pitot tube flow sensor, 191
PLC configurations, 37–41
PLC input circuits, 120
PLC power connection, 116–119
PLCs. *See* Programmable logic
        controllers
PLC sequencers, 85–86, 94–96
PLC update, 44
PLC wiring diagram, 126
PLC with common inputs, 121
PLC with isolated inputs, 122
PNP sensor load connection, 153
PNP (sourcing) output, 152
PNP units, 131, 132
Potentiometer, 204
Power rails, 13
Power supple, 40–41
Power-up scan, 76
PRESET, 83

Preset, 12
Pressure, measuring, 181–186
Pressure hammer, 182
Pressure sensitive mat switch,
    269–270
Pressure transducers, 183–185
Processor, 39
Process variable (PV), 213
Program components, 51
Programmable logic controller block
    diagram, 42
Programmable logic controllers,
    35–49
    block diagram, 42–43
    components, 39–41
    executing a ladder diagram,
        45–47
    historical history, 36–37
    PLC configurations, 37–41
Programming.
    advanced techniques, 72–102
    fundamental, 50–71
    mnemonic, 103–114
Programming unit, 41
Proportional band, 213n
Proportional gain, 213
Proportional gain amplifier, 213
Proportional gain (k_p), 231, 232
Proportional sensors, 152
Proximity sensors, 156–167
    capacitive, 159–161, 176
    inductive, 156–159
    optical, 163–167
    ultrasonic, 161–163
Pull ropes, 270
Pulse-width modulation (PWM),
    251, 256
Pulse-width modulator motor speed
    control, 251
Pushbutton, 4
Pushbutton switch actuators, 5–8
PV, 213
PWM, 251, 256

R, 15
Rails, 13
RAM memory, 42

Rate damping, 223
Red LED sensors, 163
Reference designator, 15–16
Reference designator prefixes, 15–16
Register input and output modules,
    40
Register input units, 43
Relative position, 199
Relay, 9–11
Relay contact arrangements, 127
Relay logic. See Boolean logic
Relay outputs, 127–131
Relay symbols, 10
Reset, 226
Reset line, 84
Reset rate, 231
Reset time constant, 229, 231, 233
Reset wind-up, 227
Resistance temperature device
    (RTD), 173–174
Resolution, 198
Retentive timer, 88
Retro-reflective optical sensor,
    165–167
Reversible sequencers, 86
ROM memory, 42
Rotary switch, 7
R-S flip flop, 73–74
RTD, 173–174
RUN, 32
Rung, 17
RUN switches, 24

S, 16
Safety. See System integrity and
    safety
Safety interlocks, 268–272
    interlock switches, 268–269
    light curtains, 270–272
    pressure sensitive mat switch,
        269–270
    pull ropes, 270
Saturation, 227
Scan, 47
Scan cycle, 48
Scan rate, 47
Scan time, 47

SCR, 252–255
SCR dc motor control circuit, 253
Seal-in contact, 23, 64
Sealing contact, 23, 64
Seebeck voltage, 172
Selector switch, 7
Sensors
    acceleration, 193–194
    angle position, 194–204. See also
        Angle position sensors
    defined, 150, 170
    discrete, 151
    flow, 189–191
    inclination, 192
    magnetostrictive, 206–209
    proportional, 151
    proximity, 156–167. See also
        Proximity sensors
    strain gage pressure, 183–185
    ultrasonic distance, 206
    variable reluctance pressure, 185,
        186
Sequencer, 85–86, 94–96
Setpoint (SP), 172, 213
Shadow mask, 196, 197
Shoe box, 38
Shoebox-style PLCs, 38
Shrouded actuator, 5
Silicon controlled rectifiers (SCRs),
    252–255
Simple branches, 107–110
Simple closed-loop dc motor speed
    control system, 215
Simple closed-loop systems,
    212–214
Single axis strain gage, 178
Single board PLCs, 37
Single cycle, 26–28
Single-cycle circuit, 27
Single-phasing condition, 245
Single point grounding, 146
Sinking (NPN) outputs, 131, 132,
    152
Slide potentiometer, 204
Slidewire potentiometers, 204
Slotted disk, 194–196
Solid state outputs, 131–136

Sourcing (PNP) outputs, 131, 132, 152
SP, 213
SS, 16
Stack, 108
Standard timers, 87
Stiction, 223
STO command, 105
STOP, 32
Strain, 177
Strain gage, 177–181
Strain gage accelerometers, 194
Strain gage in Wheatstone bridge circuit, 179
Strain gage pressure sensor, 183–185
Stress, 177
Strobe, 202
Switch, 4–8
    bellows, 181–182
    bi-metallic, 170–172
    cam, 8
    drag disk flow, 186–188
    E-stop, 6
    float level, 175–176
    interlock, 268–269
    limit, 7–8
    mushroom, 6–7
    pressure sensitive mat, 269–270
    pushbutton, 4
    selector, 7
Switch actuators, 5–8
Switch interlocks, 268–269
Symbols
    ASCII relay, 11
    discrete output pressure switch, 182
    discrete output temperature switch, 172
    discrete sensor, 151
    industrial electrical, 276–279
    logic, 275
    relay, 10
    TOF time delay relay, 13
    TON time delay relay, 12
System block diagram, 42–43

System integrity, 260–263
System integrity and safety, 260–273
    equipment temperature considerations, 263–264
    fail-safe wiring/programming, 264–268
    NEMA/IEC ratings, 261–263
    safety interlocks, 268–272. *See also* Safety interlocks
    system integrity, 260–263

T, 15
T_d adjustment, 231, 232
T_i adjustment, 231, 233
T_u, 235
TDRs, 11–13
Temperature measurement, 171–174
T flip flop, 79–81
Thermal dispersion flow switch, 188–189
Thermal overload, 245–247
Thermocouple, 172–173
THHN, 15
Three 1-1 phase generator output schematic, 117
Three-phase powered dc motor speed control, 255
Throttle damping, 223
Thru-beam optical sensor, 164
Tilt gages, 192
Tilt switches, 192
Time delay relays (TDRs), 11–13
Timed one shot, 93, 94
Timed sequencer, 94–96
Timer used as clock, 89, 90
TOF relay, 13
TOF time delay relays, 13
TON relay, 12
TON time delay relay, 12
TR, 16
Transducer
    defined, 170
    pressure, 183–185
    variable reluctance pressure, 185, 186
Transistor output units, 131–134

Transistor output wiring, 133
Transistor sinking output, 132
Transistor sourcing output, 132
Transitional contact, 62
Triac output unit, 134–136
Triac output wiring, 135
TTL output units, 136
Tuning constants, 231
Tuning the PID, 229–240
    adjust and observe, 231–233
    autotuning, 239–240
    open-loop method, 236–239
    oscillation method, 235–236
    ZN method, 233–239
Turbine flow sensor, 189–190
Two-handed anti-tie down, anti-repeat, single-cycle circuit, 29
Two-handed palming operation, 6
Two-handed RUN actuation, 24–26
Two-hand operation, 32
Two-rung ladder diagram, 68
Two-rung timed one shot, 94

Ultimate gain (k_u), 235
Ultimate period (T_u), 235
Ultrasonic distance sensor, 206
Ultrasonic proximity sensor, 161–163
Unipolar converter, 141, 142, 144
Unstable reading, 147
Update I/O, 44
Update-solve the ladder sequence, 43
UP/DN, 86
UP/down counter, 85
Uprights, 13
Useable sensing area, 162

Vacuum, measuring, 181–186
Variable frequency motor drive (VFD), 139, 256-258
Variable reluctance pressure sensor, 185, 186
Variable speed (frequency) ac motor drive, 255–258
VFD, 139, 256–258
Voltage resolution, 141

Washing machine program, 95
Washing machine timing chart, 94
1/4 wave decay method, 234
Wheatstone bridge, 179
Wire marker, 14
Wiring, 14–15
Wiring techniques, 115–138
    inputs having single common,
        121–124
    input wiring, 119–121

isolated inputs, 124–126
output wiring, 126–127
PLC power connection,
    116–119
relay outputs, 127–131
solid state outputs, 131–136

Ziegler, J. G., 234
Ziegler-Nichols open-loop tuning
    method, 236–239

Ziegler-Nichols oscillation tuning
    method, 235–236
Ziegler-Nichols tuning method,
    233–239
ZN tuning method, 233–239
Zone, 30–31, 96–98
Zone begin, 96, 98
Zone end, 96, 98